Inheritance of Characteristics in Domestic Fowl
Some Basic Genetics of Poultry

by Charles B. Davenport

with an introduction by Jackson Chambers

This work contains material that was originally published in 1909.

This publication is within the Public Domain.

This edition is reprinted for educational purposes and in accordance with all applicable Federal Laws.

Introduction Copyright 2018 by Jackson Chambers

The World's Largest Selection of Vintage Poultry Books

www.VintagePoultry.com

Self Reliance Books

Get more historic titles on animal and stock breeding, gardening and old fashioned skills by visiting us at:

http://selfreliancebooks.blogspot.com/

Introduction

I am pleased to present yet another title on Poultry.

The work is in the Public Domain and is re-printed here in accordance with Federal Laws.

As with all reprinted books of this age that are intended to perfectly reproduce the original edition, considerable pains and effort had to be undertaken to correct fading and sometimes outright damage to existing proofs of this title. At times, this task is quite monumental, requiring an almost total "rebuilding" of some pages from digital proofs of multiple copies. Despite this, imperfections still sometimes exist in the final proof and may detract from the visual appearance of the text.

I hope you enjoy reading this book as much as I enjoyed making it available to readers again.

Jackson Chambers

TABLE OF CONTENTS.

	PAGE
INTRODUCTION	3
CHAPTER I. THE SPLIT OR Y COMB	5
A. Interpretation of the Y Comb	5
B. Variability of the Y Comb and Inheritance of the Variations	12
CHAPTER II. POLYDACTYLISM	17
A. Types of Polydactylism	17
B. Results of Hybridization	18
CHAPTER III. SYNDACTYLISM	29
A. Statement of Problem	29
B. Results of Hybridization	32
CHAPTER IV. RUMPLESSNESS	37
CHAPTER V. WINGLESSNESS	42
CHAPTER VI. BOOTING	43
A. Types of Booting	43
B. Normal Variability	43
C. Results of Hybridization	46
CHAPTER VII. NOSTRIL-FORM	59
CHAPTER VIII. CREST	67
CHAPTER IX. COMB-LOP	69
CHAPTER X. PLUMAGE COLOR	71
A. The Gametic Composition of the Various Races	71
1. White	71
2. Black	72
3. Buff	72
B. Evidence	72
1. Silkie × Minorca (or Spanish)	72
2. Silkie × White Leghorn	75
3. Silkie × Buff Cochin	76
4. White Leghorn × Black Minorca	77
5. White Leghorn × Buff Cochin	77
6. Black Cochin × Buff Cochin	78
CHAPTER XI. INHERITANCE OF BLUE COLOR, SPANGLING, AND BARRING	79
A. Blue Color	79
B. Spangling	80
C. Barring	81
1. White Cochin × Tosa	81
2. White Leghorn Bantam × Dark Brahma	82
3. White Leghorn Bantam × Black Cochin	82
CHAPTER XII. GENERAL DISCUSSION	85
A. Relation of Heredity and Ontogeny	85
B. Dominance and Recessiveness	88
C. Potency	92
D. Reversion and the Factor Hypothesis	93
E. The Limits of Selection	94
1. Increasing the Red in the Dark Brahma × Minorca Cross	94
2. Production of a Buff Race by Selection	95
F. Non-inheritable Characters	96
G. The Rôle of Hybridization in Evolution	97
LITERATURE CITED	99

INHERITANCE OF CHARACTERISTICS IN DOMESTIC FOWL.

BY
CHARLES B. DAVENPORT.

INTRODUCTION.

A series of studies is here presented bearing on the question of dominance and its varying potency. Of these studies, that on the Y comb presents a case where relative dominance varies from perfection to entire absence, and through all intermediate grades, the average condition being a 70 per cent dominance of the median element. When dominance is relatively weak or of only intermediate grade the second generation of hybrids contains extracted pure dominants in the expected proportions of 1:2:1; but as the potency of dominance increases in the parents the proportion of offspring with the dominant (single) comb increases from 25 per cent to 50 per cent. This leads to the conclusion that, on the one hand, dominance varies quantitatively and, on the other, that the degree of dominance is inheritable.

The studies on polydactylism reveal a similar variation of potency in dominance and show, in Houdans at least, an inheritance of potency (table 11), and moreover they suggest a criticism of Castle's conclusion of inheritance of the degree of polydactylism.

Syndactylism illustrates another step in the series of decreasing potency of the dominant. On not one of the F_1 generation was the dominant (syndactyl) condition observed; and when these hybrids were mated together the dominant character appeared not in 75 per cent but in from 10 per cent to 0 per cent of the offspring. The question may well be asked: What is then the criterion of dominance? The reply is elaborated to the effect that, since dominance is due to the presence of a character and recessiveness to its absence, dominance may fail to develop, but recessiveness never can do so. Consequently two extracted recessives mated *inter se* can not throw the dominant condition; but two imperfect dominants, even though indistinguishable from recessives, will throw dominants. On the other hand, owing to the very fact that the dominant condition often fails of development, two extracted "pure" dominants will, probably always, throw some apparent "recessives." Now, two syndactyls have not been found that fail (in large families) to throw normals, but extracted normals have been found which, bred *inter se*, throw only normals; hence, "normal-toe" is recessive. In this character, then, dominance almost always fails to show itself in the heterozygote and often fails in pure dominants.

The series of diminishing potency has now brought us to a point where we can interpret a case of great difficulty, namely, a case of rumplessness. Here a dominant condition was originally mistaken for a recessive condition, because it never fully showed itself in F_1 and F_2. Nevertheless, in related individuals, the condition is fully dominant. We thus get the notion that a factor that normally tends to the development of a character may, although present, fail to develop the character. Dominance is lacking through *impotence*.

The last term of the series is seen in the wingless cock which left no wingless offspring in the F_1 and F_2 generations. In comparison with the results gained with the rumpless cock, winglessness in this strain is probably dominant but impotent.

When a character, instead of being simply present or absent, is capable of infinite gradations, inheritance seems often to be blending and without segregation. Two cases of this sort—booting and nostril-height—are examined, and by the aid of the principle of imperfect dominance the apparent blending is shown to follow the principle of segregation. Booting is controlled by a dominant inhibiting factor that varies greatly in potency, and nostril-height is controlled by an inhibiting factor that stops the overgrowth of the nasal flap which produces the narrow nostril.

The extracted dominants show great variability in their progeny, but the extracted recessives show practically none. This is because a positive character may fail to develop; but an absent character can not develop even a little way. The difference in variability of the offspring of two extracted recessives and two extracted dominants is the best criterion by which they may be distinguished, or by which the *presence* (as opposed to the absence) of a factor may be determined.

The crest of fowl receives especial attention as an example of a character previously regarded as simple but now known to comprise two and probably more factors—a factor for erectness, one for growth, and probably one or more that determine the restriction or extension of the crested area.

The direction of lop of the single comb is an interesting example of a character that seems to be undetermined by heredity. In this it agrees with numerous right and left handed characters. It is not improbable that the character is determined by a *complex* of causes, so that many independent factors are involved.

A series of studies is presented on the inheritance of plumage color. It is shown that each type of bird has a gametic formula that is constant for the type and which can be used with success to predict the outcome of particular combinations. New combinations of color and "reversions" receive an easy explanation by the use of these factors. The cases of blue, spangled, and barred fowl are shown also to contain mottling or spangling factors.

CHAPTER I.

THE SPLIT OR Y COMB.

A. INTERPRETATION OF THE Y COMB.

When a bird with a single comb, which may be conveniently symbolized as I, is crossed with a bird with a "V" comb such as is seen in the Polish race, and may be symbolized as oo, the product is a split or Y comb. This Y comb is a *new form*. As we do not expect new forms to appear in hybridization, the question arises, How is this Y comb to be interpreted? Three interpretations seem possible. According to one, the antagonistic characters (allelomorphs) are I comb and oo comb, and in the product neither is recessive, but both dominant. The result is a case of particulate inheritance—the single comb being inherited anteriorly and the oo comb posteriorly. On this interpretation the result is not at all Mendelian.

According to the second interpretation the hereditary units are not what appear on the surface, but each type of comb contains two factors, of which (in each case) one is positive and the other negative. In the case of the I comb the factors are presence of median element and absence of lateral or paired element; and in the case of the oo comb the factors are absence of median element and presence of lateral element. On this hypothesis the two positive factors are dominant and the two negative factors are recessive.

The third hypothesis is intermediate between the others. According to it the germ-cells of the single-combed bird contain a median unit character which is absent in the germ-cells of the Polish or Houdan fowl. This hypothesis supposes further that the absence of the median element is accompanied by a fluctuating quantity of lateral cere, the so-called V comb.

The split comb is obtained whenever the oo comb is crossed with a type containing the median element. Thus, the offspring of a oo comb and a pea comb is a split pea comb, and the offspring of a oo comb and a rose comb is a split rose. The three hypotheses may consequently be tested in three cases where a split comb is produced.

TABLE 1.

	I	Y	No median.
I × I	100	0	0
I × Y	50	50	0
I × no median	0	100	0
Y × no median	0	50	50
No median × no median	0	0	100

The first and third hypotheses will give the same statistical result, namely, the products of two Y-combed individuals of F_1 used as parents, will exhibit the following proportions: median element, 25 per cent; split comb, 50 per cent; and no median element, 25 per cent. These proportions will show themselves, whatever the generation to which the Y-combed parents belong, whether both are of generation F_1, or F_2, or F_3, or one parent of one generation and the other of another. Other combinations of parental characters should give the proportions in the progeny shown in table 1.

On the second hypothesis, on the other hand, the proportions of the different kinds occurring in the progeny will vary with the generation of the parents. This hypothesis assumes the existence in each germ-cell of the original parent of two comb allelomorphs, M and l in single-combed birds and m and L in the Polish fowl, the capital letter standing for the presence of a character (Median element or Lateral element) and the small letter for the absence of that character. Consequently, after mating, the zygote of F_1 contains all 4 factors, $MmLl$, and the soma has a Y comb; but in the germ-cells, which contain each only 2 unlike factors, these factors occur in the following 4 combinations, so that there are now 4 kinds of germ-cells instead of the 2 with which we started. These are ML, Ml, mL, and ml. Furthermore, since in promiscuous mating of birds these germ-cells unite in pairs in a wholly random fashion, 16 combinations are possible, giving 16 F_2 zygotes (not all different) as shown in table 2.

TABLE 2.

Type.	Zygotic constitution.	Soma.	Type.	Zygotic constitution.	Soma.	Type.	Zygotic constitution.	Soma.
a	M_2L_2*	Y	g	$MmLl$	Y	n	$mlML$	Y
b	M_2Ll	Y	h	Mml_2	I	o	$mlMl$	I
c	MmL_2	Y	i	$mLML$	Y	p	m_2Ll	oo
d	$MmLl$	Y	k	$mLMl$	Y	q	m_2l_2	Absent
e	M_2Ll	Y	l	m_2L_2	oo			
f	M_2l_2	I	m	m_2Ll	oo			

*This convenient form of zygotic formula, using a subscript 2 instead of doubling the letter, is proposed by Prof. W. E. Castle.

It is a consequence of this second hypothesis that, in F_2, of every 16 young 9 should have the Y comb; 3 the I comb; 3 the oo comb, and 1 no comb at all. It follows further that the progeny of two F_2 parents will differ in different families. Thus if a Y-combed bird of type a be mated with a bird of any type, all of the progeny will have the Y comb.

From Y-combed parents of various types taken at random 4 kinds of families will arise having the following percentage distribution of the different types of comb:

1. Y comb, 100 per cent.
2. Y comb, 75 per cent; I comb, 25 per cent.
3. Y comb, 75 per cent; oo comb, 25 per cent.
4. Y comb, 56.25 per cent; I comb, 18.75 per cent; oo comb, 18.75 per cent; absent, 6.25 per cent.

Again, mating two extracted I combs of F_2 should yield, in F_3, two types of families in equal frequency as follows:

1. I comb, 100 per cent.
2. I comb, 75 per cent; no comb, 25 per cent.

Again, mating two extracted oo combs of F_1 should yield, in F_2, two types of families in equal frequency, as follows:

1. oo comb, 100 per cent.
2. oo comb, 75 per cent; no comb, 25 per cent.

Single comb × Y comb should give families of the types:

1. Y comb, 100 per cent.
2. Y comb, 50 per cent; I comb, 50 per cent.
3. Y comb, 50 per cent; oo comb, 50 per cent.
4. Y comb, 25 per cent; I comb, 25 per cent; oo comb, 25 per cent; absent, 25 per cent.

Mating oo comb and Y comb should give the family types:

1. Y comb, 100 per cent.
2. Y comb, 50 per cent; oo comb, 50 per cent.
3. Y comb, 50 per cent; I comb, 50 per cent.
4. Y comb, 25 per cent; oo comb, 25 per cent; I comb, 25 per cent; no comb, 25 per cent.

Finally, I comb and oo comb should give the following types of families:

1. Y comb, 100 per cent.
2. I comb, 100 per cent.
3. Y comb, 50 per cent; oo comb, 50 per cent.
4. I comb, 50 per cent; no comb, 50 per cent.

Now, what do the facts say as to the relative value of these three hypotheses? Abundant statistics give a clear answer. In the first place, the progeny of two Y-combed F_1 parents is found to show the following distribution of comb types: Y comb 471, or 47.3 per cent; I comb 289, or 29.0 per cent; oo comb 226, or 22.7 per cent; and no comb 10, or 1 per cent. The presence of no comb in F_2 speaks for the second hypothesis, but instead of the 6.25 per cent combless expected on that hypothesis only 1 per cent appears. There is no close accord with expectation on the second hypothesis.

Coming now to the F_3 progeny of two Y-combed parents, we get the distribution of families shown in table 3.

TABLE 3.

Pen No.	Parents.		Comb in offspring.			
	♀ (F_2).	♂ (F_2)	I	Y	oo	Absent.
707	366	1378	18	16	9	....
	522	1378	1	1	0	....
763	2250	2247	9	5	4	1
	2700	2247	3	5	3	1
	3799	2247	5	4	3	....
769	1305	911	7	4	6	....
	2254	911	15	15	7	....
Totals (142)........			58	50	32	2
Proportions (per cent)			40.8	35.2	22.5	1.4
					23.9	

An examination of these families shows not one composed exclusively of Y-combed individuals nor those (of significant size) containing Y-combed and I-combed or oo-combed individuals exclusively, much less in the precise proportion of 3 : 1, yet such should be the commonest families if the second hypothesis were true. Notwithstanding the marked deviation—to be discussed later—from the expected proportions of I, 25 per cent; V, 50 per cent; oo, 25 per cent, the result accords better with the first or third hypothesis. Since on either of these hypotheses the same proportions of the various types of comb are to be expected in the progeny of V-combed parents of whatever generation, it is worth recording that from such parents belonging to all generations except the first the results given in table 4 were obtained, and it will be noticed that these results approach expectation on the first or third hypothesis.

TABLE 4.

	I	V	oo	Absent.	Total.
Frequency	235	291	144	12	682
Percentage	34.5	42.7	21.1	1.8	....

The progeny of two extracted single-combed parents of the F_2 generation give in 3 families the following totals: Of 95 F_3 offspring, 94 have single combs; one was recorded from an unhatched chick as having a *slightly* split comb, but this was probably a single comb with a slight side-spur, a form that is associated with purely I-combed germ-cells. This result is in perfect accord with the second and third hypotheses, but is irreconcilable with the first hypothesis.

The progeny of two extracted oo-combed parents is given in table 5.

TABLE 5.

Pen No.	Parents.		Comb in offspring.				Pen No.	Parents.		Comb in offspring.			
	♀ (F_2).	♂ (F_2).	I	V	oo	Absent.		♀ (F_2).	♂ (F_2).	I	V	oo	Absent.
729	2255	936	....	*4	36	....	820	2016	4731	....	....	10	....
	2269	936	....	....	29	....		2255	4731	....	....	16	....
	369	1390	1	....	3	....		5143	4731	....	....	45	....
756	1067	1390	....	....	8	1		6479	4731	....	....	31	....
	1113	1390	....	....	13	4		†2618	5119	†1	....	23	....
	2011	444	....	....	10	....		3776	5119	....	....	28	....
	2011	2621	....	....	9	....		4404	5119	....	....	9	....
	2333	444	....	*5	11	....	832	4732	5119	....	....	3	....
	2333	2621	....	*1	2	....		5803	5119	....	....	21	2
762	2618	444	....	....	2	....		6481	5119	....	....	11	....
	2618	2621	....	....	5	....	834	2324	5090	....	....	26	....
	3776	444	....	....	2	....	Total			2	11	367	7
	3776	2621	....	1	14	....							

* Median element recorded as "small" in these offspring. † A median element visible in the mother, No. 2618.

The distribution of offspring in the 24 families of table 5 is in fair accord with any of the three hypotheses, but seems to favor the second, for that hypothesis calls for families with combless children, whereas such are not to be expected on the first hypothesis. Moreover, agreement with the second hypothesis is fairly close, for that calls for 3 families with combless children and there were actually 3 such families showing a total of 1.8 per cent combless, where expectation is 2.8 per cent. What is opposed to any hypothesis is the appearance of some Y-combed offspring; and to account for this the hypothesis is suggested that the germ-cells of some parents with oo comb contain traces of the I-comb determiner. The word "traces" is used because the median element in these Y-combed offspring is practically always very small. It is fair, consequently, to conclude that oo×oo gives oo-combed, and occasionally combless, offspring. This conclusion is further supported by the statistics derived from extracted oo comb of *all* generations bred *inter se*, which give: Y 11, oo 427, and no comb 8, where the 11 Y-combed birds are those just referred to as progeny of F_2 parents. The non-median comb, consequently, probably contains only non-median germ-cells.

TABLE 6.

Pen No.	Parents.						Offspring.		
	♀ (F_2).	Form of comb.	Degree of splitting.	♂ (F_2).	Form of comb.	Degree of splitting.	I	Y	oo
			P. ct.			P. ct.			
628	427	Y	5	439	I	0	5	1	....
	722	Y	20	439	I	0	1	5	....
	725	Y	10	439	I	0	5	3	....
629	427	I	0	491	Y	50	9	6	....
765	1790	I	0	1794	Y	90	17	25	....
802	3846	I	0	6652	Y	90	8	5	....
	5025	I	0	6652	Y	90	14	11	2
	5087	I	0	6652	Y	90	13	17	2
812	4254	I	0	4118	Y	90	15	13	....
	5540	I	0	4118	Y	90	8	9	....
Totals (189)							95	95	4
Percentages							49.0	49.0	2.0

The mating of extracted I comb and Y comb, both of the second (or later) hybrid generation, gives the following distribution of types in the offspring (table 6): Y comb 95 (49 per cent); I comb 95 (49 per cent); oo comb 4 (2 per cent). In detail the results given in table 6 accord badly with the second hypothesis, which demands some families with 100 per cent Y comb.

The mating of extracted oo comb×Y comb, where both parents are of the second hybrid generation, gave the distribution of comb types in the 6 families that are recorded in table 7.

TABLE 7.

Pen No.	Parents.		Offspring.			
	♀ (F_2).	♂ (F_2).	I	Y	oo	Absent.
634	298	444	0	15	18	....
	366	444	5	23	15	....
729	913	936	2	28	37	....
	935	936	....	13	39	....
756	1043	1390	....	13	11	1
	1048	1390	....	0	5	....
Totals (214)			7	92	115	1

The single comb recorded in the case of 7 birds is doubtless merely the limiting condition of a Y comb in which the median element is developed to its fullest extent. All but 2 of the 7 were recorded from early embryos when an incipient bifurcation would be more difficult to detect. This explanation applies generally, and accounts for the usual excess of I comb when compared with Y comb, as for instance in table 3, page 7. Returning to table 7, it is, consequently, probable that only the Y-combed and non-median-combed offspring are produced and that they are in the proportion of 99 to 115 or of 46 per cent to 54 per cent. If we add together all records of a oo×Y cross, disregarding the generation of the parents, we get a total I comb 5,* Y comb 177, oo comb 172, and absent 3, or 182 (51 per cent) with the median element and 175 (49 per cent) without. Thus the oo×Y cross gives the 1 : 1 proportion called for on the first and third hypotheses and not at all the variety required by the second hypothesis.

TABLE 8.

Pen No.	Mother.			Father.		Comb in offspring.			
	No.	Comb.	P. ct. split.	No.	Comb.	I	Y	oo	Abs.
704	65 F$_1$	Y	50	1420 F$_2$	Absent	....	10	6	8
	1061 F$_2$	Y	50	1420 F$_2$	...Do...	....	4	....	1
819	57 F$_1$	Y	50	1420 F$_2$	...Do...	....	8	6	5
	65 F$_1$	Y	60	1420 F$_2$	...Do...	....	1	....	1
Total						0	23	12	15

Finally, we must consider the result of mating a bird without papillæ (No. 1420, pen 704) with a median-combed hen (480). When this typical single-combed hen was used the 49 progeny were all of the Y type.† This proves that the combless type behaves only as an extreme of the non-median type.

When Y-combed hens were used with the combless cock the offspring had Y comb and non-median-comb in nearly equal numbers, 23 : 27 (table 8), but the latter included an unusually large proportion of combless fowl (15 in 27). When a combless hen (No. 4257) was used, 9 of the offspring had oo comb and 2 no comb; not a greater proportion of combless birds than in the no-comb×Y-combed cross. All of these facts indicate that "combessness" is not entire absence of the comb factors, but a minimum case of the oo or paired comb. This result is opposed to the second hypothesis.

The statistics of all matings between I, Y, and no comb on the one side and no comb on the other thus speak unanimously for the conclusion that in these matings we are not dealing with 2 pairs of allelomorphs, but with a single comb and its absence (third hypothesis) or with a case of

---

* Excluding 6 doubtful because from too young embryos and not observed by myself.
† One is reported as having a I comb; probably the limiting condition, again.

particulate inheritance (first hypothesis). Moreover, it must be said that the split comb is obtained also when the Polish-Houdan comb is crossed with a pea comb or a rose comb; and the pea and rose combs can not be said to have "lateral comb absent," as required by the second hypothesis. Consequently the second hypothesis is definitely excluded.

It now remains to decide between the two remaining hypotheses. First of all, it may be said that the perfection with which I and oo combs can be extracted from Y-combed birds indicates that we are here dealing with a case of Mendelian inheritance and, in so far, favors the third hypothesis. To accord with the theory of particulate inheritance, of which the first hypothesis is a special case, the two united characters should transmit the mosaic purely; but this they do not do. Hence the third hypothesis is to be preferred to the first.

Comblessness is a necessary consequence of the second hypothesis and is inexplicable on the first hypothesis. On the third hypothesis it may be accounted for as follows: Absence of single comb is allelomorphic to its presence. The lateral comb is a character common to fowl either with or without the median comb, but it is ordinarily repressed in the birds with single comb and gains a large size when the median element is absent. It is a very variable element. At one extreme it forms the cup comb; at the other there is an absence of any trace of comb. My own records show all grades between these extremes, including minute papillæ on both sides of the head or on one side only, low paired ridges, the butterfly comb, and cup comb shorter than normal. This variability of the lateral element is comparable to the fluctuation in size of the single comb itself, as illustrated by the Single-comb Minorca on the one hand and the Cochin on the other. It is comparable, also, to the fluctuation in the paired part of the Y comb, which we shall consider in the next section, and to the variability of the oo comb as met with in the pens of fanciers.

The foregoing considerations do not, at first sight, account for the Y comb as seen in F_1. Yet they provide us with all the data for an explanation. Median comb of the Minorca dominates over no median of the Polish, and so in F_1 we have the median element represented. But, on the well-known principle of imperfection of dominance in F_1, the median comb is usually incomplete and, probably for some ontogenetic reason, incomplete only behind. The incompleteness behind permits the development there of the elsewhere repressed lateral comb, and we therefore have the Y comb— evidence at the same time of a repressed lateral-comb Anlage in the single-combed birds and of imperfection of dominance of the single comb in the first hybrid generation.

B. VARIABILITY OF THE Y COMB AND INHERITANCE OF THE VARIATIONS.

As already stated, the proportions of the median and the lateral elements in the Y comb are very variable; the median element may, indeed, constitute anywhere from 100 per cent to 0 per cent of the entire comb. Even full brothers and sisters show this variability. Thus the offspring of No. 13 ♀ Single-comb Minorca and No. 3 ♂ Polish have the median element of the Y comb ranging from 0 per cent to 70 per cent of the whole comb. Notwithstanding this variability of the median element in any family there is a difference in the average and the range of variability in families where different races are employed. Thus the offspring of two Polish×Minorca

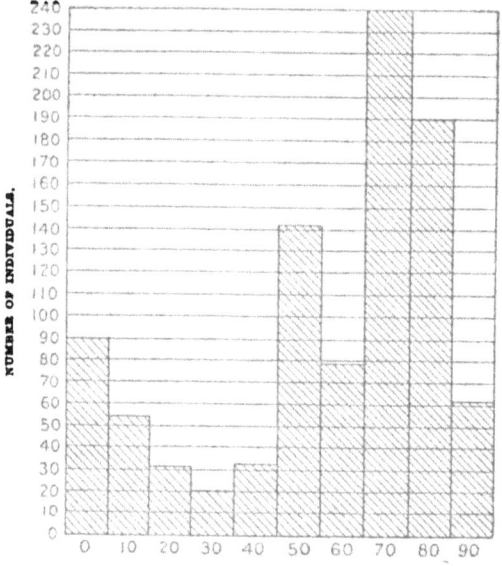

Fig A.—The frequency of the different forms of Y comb, each form being based on the percentage of the median element of the Y comb to the entire length of comb.

crosses show an *average* of 46 per cent of the median element in the comb; the Houdan×Minorca cross gives combs with 60 per cent of the median element; and in the combs of the offspring of two Houdan×White Leghorn crosses there is, on the average, 71 per cent of the median element. The Houdan×Dark Brahma (pea comb) gives combs with an average of 87 per cent median element and the Polish×Rose-comb Minorca cross gives 89 per cent median. The rose-combed hens used in this last cross were heterozygous, having single comb recessive; consequently they produced also chicks with typical Y combs. Such had, on the average, only 59 per cent of the median element and were thus in striking contrast with the slightly split rose combs. In the case of the partially split rose combs the median element ranged from 60 per cent to 100 per cent of the whole length of the comb; but in the split single combs the range is from 0 to 100 per cent. Thus, in the two cases, the proportion of the median element and the range of its variability differ greatly.

Also, in generations subsequent to the first, the Y comb exhibits this same variability. We have already seen that the progeny of the Y-combed offspring of any generation may be compared with those of any other, and so we may mass together the progeny of all hybrid generations so long as they are derived from the same ancestral pure races.

In inquiring into the meaning of this variability we must first construct the polygon of frequency of the various grades of median element. This is plotted in fig. A, which is a composite whose elements are, however, quite like the total curve. There is one empirical mode at 70 per cent and another at 0 per cent. The smaller mode at 50 per cent is, I suspect, due to the tendency to estimate in round numbers, and may be, in this discussion, neglected. From this polygon we draw the conclusions, first, that the median element in the Y comb tends to dominate strongly over the absence of this element, as 7 : 3, and, second, that dominance is rarely complete. Yet there is an important number of cases, even in F_1, where the median element is almost or completely repressed (down to 10 to 0 per cent of the whole) and the comb consists of two high and long lateral elements—the "cup comb" of Darwin. There are, then, in the offspring of a median-combed and a non-median-combed parent, two types with few intergrades—the type of slightly incomplete dominance of the median element and the type of very incomplete dominance.

We have now to consider how these two types of comb and their fluctuations behave in heredity. When two parents having each combs of the 70 per cent or 80 per cent median type are mated, their offspring belong to the three categories of I, Y, and "no-median" comb, but the relative frequency of these three categories is not close to the ideal of 25 per cent, 50 per cent, and 25 per cent, respectively. For there is actually in 336 offspring a marked excess of the I comb, 36 per cent, 44 per cent, and 20 per cent, respectively, resulting. When, on the other hand, two parents having each combs of the 10 per cent and 0 per cent types are mated their offspring are of the same three categories and the proportions actually found in 241 offspring (28 per cent, 47 per cent, 25 per cent) closely approximate the ideal. It is clear, then, that even the cup comb, without visible median element, has such an element in its germ-cells and is totally different in its hereditary behavior from the Polish comb, in which the median element is absent, not only from the soma, but also from the germ-cells.

We have seen in the last paragraph that the Y comb with only 10 per cent to 0 per cent median element has germ-cells bearing median comb as truly as the Y comb containing 70 per cent to 80 per cent median element, but we have also seen that in the latter case there is an excess of single-combed progeny. We have now to inquire whether, in general, there is a close relation between the proportion of median element in the comb of the parents and the percentage of single-combed offspring. These relations are brought out in the lower half of table 9.

TABLE 9.—*Frequency of the different proportions of single element in the combs of offspring of parents having the average proportion of median element given in the column at the left.*

	Y combs.										
	Offspring.										
	0	10	20	30	40	50	60	70	80	90	Total.
Parents 0....	21	5	4	3	4	6	5	10	8	1	67
10....	21	5	3	0	3	9	2	4	2	0	49
20....	5	4	2	1	0	4	2	12	0	1	31
30....	8	17	8	10	9	22	12	30	8	3	127
40....	9	7	4	2	7	39	18	46	26	5	163
50....	7	5	2	1	5	32	13	48	35	11	159
60....	10	7	2	2	2	19	14	47	51	15	169
70....	9	2	4	0	1	6	7	28	41	11	109
80....			1	1	1	1	6	12	11	6	39
90....		2	1	0	0	3	0	3	8	9	26
Total.....	90	54	31	20	32	141	79	240	190	62	939

	All types of combs in offspring.						
	Number of offspring.	I		Y		Non-median.	
		No.	P. ct.	No.	P. ct.	No.	P. ct.
Parents 0....	146	42	29	67	46	37	25
10....	99	25	25	49	50	25	25
20....	73	22	30	31	43	20	27
30....	249	61	25	127	51	61	24
40....	309	73	24	163	53	73	23
50....	329	93	28	159	48	77	23
60....	368	120	33	169	46	79	21
70....	232	80	35	109	47	43	18
80....	104	42	40	39	38	23	22
90....	75	38	51	26	34	11	15
Total.....	1984	596	30.0	939	47.3	449	22.7

The proportion of single-combed offspring in the total filial population is 30.0 per cent, a departure of such magnitude from the expected 25 per cent as to arrest our attention. Further inspection of table 9 shows that the excess of single-combed offspring is found only in the lower half of the series. When the percentage of median element in the parents is under 50 the proportions of I, Y, and no-median combs are as 25.5 per cent, 49.8 per cent, 24.7 per cent, or close to expectation; but when the percentage is 50 or over the proportions are, on the average, 33.6 per cent, 45.2 per cent, and 21.2 per cent, a wide departure from expectation, 1108 individuals being involved. An examination of table 9 shows, moreover, that the proportion of offspring with single comb rises steadily as the proportion of the median element in the parentage increases from 50 per cent. The meaning of this fact is at present obscure, but the suspicion is awakened that, while the "cup comb" and the more deeply split combs are typical heterozygotes the slightly split combs are a complex of 2 or more units, one of which is "single comb." But that this is not the explanation follows for two reasons: first, that even in the F_1 generation slightly split combs are obtained, and, second, that the offspring of the cup combs are much more variable than those of slightly split combs (70 to 90 per cent median). What is strikingly true is that, from 50 per cent up, as the proportion of the median element in the parents increases the percentage of single-combed offspring rises.

The matter may be looked at in another light. Median comb is dominant over its absence. Typically, we should expect F_1 to show a single comb; the Y comb that we actually get is a heterozygous condition due to the failure of the median comb to dominate completely. Typically we should expect F_2 to reveal 75 per cent single combs, of which 1 in 3 is homozygous and 2 in 3 are heterozygous. Owing to the failure of single comb always to dominate completely in the heterozygotes, we expect to find some of the 75 per cent with the Y comb. When in the parents dominance has been very incomplete in the heterozygote (as is the case in the 0 per cent to 40 per cent median-combed parents) we find it so in the offspring also and all heterozygotes show a Y comb of some type. But when in the parents dominance has been strong in the heterozygote (50 per cent to 90 per cent) it is so in the offspring also and only a part of the heterozygotes show the Y comb; the others show the single comb and thus swell the numbers of the single-combed type. The only objection to this explanation is found in the reduction in the percentages of the no-median type. Thus, adding together the homozygous and heterozygous median-combed offspring and comparing with the non-median-combed, we find these ratios:

Parental per cent	0–40	50	60	70	80	90
Ratio	75.3 : 24.7	76 : 23	79 : 21	82 : 18	78 : 22	85 : 15

There is a great deviation from 25 per cent in the "non-median" offspring of the 90 per cent parents, but in this particular case the total number of offspring is not large, and the deviation has a greater chance of being accidental. Altogether this explanation of the varying per cents of single comb on the ground of inheritance of varying potency in dominance seems best to fit the facts of the case.

From the foregoing facts and considerations we may conclude that the Y comb represents imperfect dominance of median over no-median comb; that there is a fluctuation in the potency of the dominance, so that the proportion of the median element varies from 0 to over 90 per cent; that the more potent the dominance of median element is in any parents the more complete will be the dominance in the offspring and the smaller will be the percentage of imperfectly dominant, or Y-combed, offspring. *Dominance varies quantitatively and the degree of dominance is inheritable.*

The index of heredity may be readily obtained in the familiar biometric fashion from table 9. This I have calculated and found to be 0.301 ± 0.002. This agrees with Pearson's theoretical coefficient of correlation between offspring and parent. The index is larger than it would otherwise be because it is measured with an *average* of the parents and these parents assortatively mated. But this instance is, in any case, an interesting example of strong inheritance of a quantitative variation.

What, it may be asked, is the relation of these facts to the general principle that inheritance is through the gametes? Why, when a gamete with the median element unites with a gamete without that element, does the zygote develop a soma that in some cases shows a nine-tenths median and sometimes a one-tenth median element? We have seen that the Y comb is a heterozygous form due to imperfection of dominance of the median element; but why this variation in the perfection of the median element? This is probably a piece of the question, why any dominance at all. We find, in general, that the determiner of a well-developed organ dominates in the zygote over the determiner of a slightly developed condition of that organ or its obsolete condition. It is as though there were in the zygote an interaction between the strong and the weak form of the determiner, and the strong won; but sometimes the victory is imperfect. In the specific case of comb the interaction between median and no-median leads to a modification, weakening, or imperfection of the median element, and this weakening varies in degree. Sometimes the weakening is inappreciable—when the comb is essentially single; sometimes it is great, and the result is a comb in which the median element is reduced to one-half; sometimes, finally, the determiner of median comb is so completely weakened by its dilution with "no-median" as not to be able to develop, and we have the cup comb with only a trace of the median element. Nevertheless, such a cup comb is heterozygous and produces both single-combed and Polish-combed germ-cells. Thus the variation in the extent of the median comb seems to point to variations in relative potency of the median comb over its absence.

CHAPTER II.

POLYDACTYLISM.

The possession of extra toes is a character that crops out again and again among the higher, typically 5-toed vertebrates. Many cases have been cited in works on human and mammalian teratology (cf. Bateson, 1904, and Schwalbe, 1906), and it is recognized that this abnormality is very strongly inherited in man. Bateson and Saunders, and Punnett (1902 and 1905), Hurst (1905), and Barfurth (1908), as well as myself in my earlier report, have demonstrated the inheritableness of the character in poultry. Bateson and Punnett (1905, p. 114) say: "The normal foot, though commonly recessive, may sometimes *dominate* over the extra-toe character, and this heterozygote may give equality when bred with recessives, just as if it were an ordinary DR." Altogether, the inheritance of extra-toe diverges so far from typical Mendelian results as to deserve further study.

A. TYPES OF POLYDACTYLISM.

There are two main types of polydactylism: that in which the inner toe (I) of the normal foot is replaced by 2 simple toes, and that in which it is replaced by two toes, of which the mediad is simple and the laterad is divided distally. The former type is characteristic of the Houdans; the latter is usually associated with the Silkies. Both conditions are, however, found in both races. The simplest condition is seen in many Houdans of my strain. It consists of 2 equal, medium-sized toes (I' and I") lying close together and parallel to or slightly convex towards each other. This condition indicates that the 2 toes, together, are to be regarded as the equivalent of the normal single toe occupying the same position. The 2 toes are, I conjecture, derived from the single toe by splitting. The first series of changes consists of the increase in length of the lateral element (I") and a corresponding decrease of the median element (I'). In the last term of the series there are only 4 toes on the foot, but the inner toe is not like the normal inner toe of poultry, but is a much elongated I".

In the Silkie, also, the series begins with 2 small, closely-applied toes (I' and I"). But when there are only 2 toes the lateral one is usually much the larger. Typically this lateral toe is, as stated, split, so that the nail is double, and the degree of splitting is variable, in extreme cases involving half or more than half of the toe. A second series of changes consists of the gradual reduction of toe I' (often concomitantly with an increase in I") which may end in its entire disappearance and thus reduce the number of toes to 5, but these are not equivalent to the 5 toes of the Houdans, since the extra Houdan toes are I′, I″, and those of the reduced Silkie are I″a and I″b. Finally, in Silkies, the inner toe (I′) may split (more or less completely), and thus the 7-toed condition arises. Moreover, in Houdans I have on one or two occasions found the lateral element (I″) bifid distally, resembling perfectly the typical condition found in the Silkies.

A simple nomenclature is suggested for these various types of extra-toes. The simple double-toed condition, as found commonly in Houdans, may be called the *duplex* type (D). The loss of I′ gives the *reduced duplex* (D′). The case of split I″, as commonly seen in the Silkie, is the *triplex* type (T); with the loss of I′ this becomes the *reduced triplex* (T′, not duplex!). The 7-toed condition of Silkies may be called the *quadruplex type* (Q); the combination split I′ and single I″ gives the *reduced quadruplex* (Q′).*

The reduction that leads to the loss of I′ consists of a loss of phalanges, as Bateson (1904) has already pointed out. It seems probable that the reduction affects first the proximal phalanges, since the distal nail-bearing phalanx is the last to disappear.

B. RESULTS OF HYBRIDIZATION.

First let us consider the result of mating extra-toed individuals belonging to "pure" extra-toed races. A typical Houdan cock (D type), of the well-known Petersen strain, was mated with 3 hens bred by me, but derived, several generations before, from the same strain. With the first hen he got 29 chicks, all with the extra-toe except one (3.3 per cent) that had 4 toes on both feet and two that had 4 toes on one foot and 5 on the other, *i. e.*, one foot simplex and one duplex. With the second he got 12 chicks, of which one had 4–5 (D) toes. The third, in 26 young, gave one with 4 toes on each foot. Thus, in 67 chicks altogether there were 2, or 3 per cent, with the normal number of toes on both feet (4–4). Unfortunately these birds did not survive, so it is not known whether they would have thrown as large a proportion of extra-toed offspring as 5-toed Houdans. Bateson's Dorkings gave about 4 per cent of 4-toed offspring. Of the 83 offspring of 6-toed Silkies, 3, or 3.6 per cent, had 4 toes on each foot. Even in pure-bred polydactyl races, consequently, the character "extra-toe" does not uniformly appear in the offspring.

Let us consider next what happens when a polydactyl individual is crossed with a normal individual. Table 10 gives the results of all matings of this sort and its most obvious result is that the polydactyl condition reappears in every family, but not, as in typically Mendelian cases, in *all* of the offspring; at least this is true of the Houdan crosses. In the Silkie crosses the 6 offspring given as having the single thumb may possibly have been of the type D′, as that type was not in mind at the time of making the record and was not always distinguished from type S. It is also clear that the offspring of Silkie crosses are more apt to be polydactyl than those of

*B. ♂., Pen 813, 935 ♀, embryo from egg of May 13.

TABLE 10.—*Frequency of the various types of toes in the first hybrid generation between a normal and an extra-toed parent.*

A. HOUDAN CROSSES.

Pen No.	Mother.			Father.			Offspring.			
	No.	Race involved.	No. of toes.	No.	Race involved.	No. of toes.	Types of toes.			Average.
							4-4	4-5	5-5	
504	8 or 11	Houdan	5-5	73	Wh. Leghorn	4-4	0	1	8	9.9
504	8	...Do...	5-5				1	3	8	9.6
	11	...Do...	5-5				2	2	7	9.5
525	8 or 11	...Do...	5-5	27	Minorca	4-4	8	3	13	9.2
727	"Y"	Dk. Brahma	4-4	831	Houdan	5-5	3	2	5	9.2
	121	...Do...	4-4				13	9	18	9.1
504	10-12	Wh. Leghorn	4-4	9	...Do...	5-5	3	2	0	8.4
	Total (110)						30	21	59	9.26
	Percentages						27.3	19.1	53.6	

B. SILKIE CROSSES.

Pen No.	Mother.			Father.			Offspring.										Average.
	No.	Race.	No. of toes.	No.	Race.	No of toes.	Types of toes.†										
							ss.	sd'.	sd.	d'd'.	d'd.	dd.	st'.	d't'.	dt'.	t't'.	
851	1002	Cochin	4-4	7526	Silkie	6-6		1		1	2			2	3		10.78
851	3410	...Do...	4-4	7526	...Do...	6-6	1?				2	7			1	3	10.43
815	131	...Do...	4-4	774	...Do...	6-6				1		8		1	1	1	10.33
851	2073	...Do...	4-4	7526	...Do...	6-6						7	1			1	10.33
734	841	...Do...	4-4	774	...Do...	6-6						3			1		10.25
851	338	...Do...	4-4	7526	...Do...	6-6			1	1		11				3	10.25
851	2299	...Do...	4-4	7526	...Do...	6-6			1?	1		4				1	10.14
851	5567	...Do...	4-4	7526	...Do...	6-6					1	10	1		1		10.08
734	840	...Do...	4-4	7526	...Do...	6-6					1	7					10.00
734	1002	...Do...	4-4	774	...Do...	6-6					2	8					10.00
851	840	...Do...	4-4	7526	...Do...	6-6						4					10.00
851	841	...Do...	4-4	7526	...Do...	6-6					1	1					10.00
744	777	Silkie	*5-6	1176	Wh. Leghorn	4-4						6					10.00
744	496	...Do...	6-6	1176	...Do...	4-4	1?					12			1		9.93
851	6956	Cochin	4-4	7526	Silkie	6-6	4?	1		2		3					9.50
	Total (138)						6	1	3	6	7	93	2	1	7	12	10.13

*Of the reduced triplex type (t'). †s, means type of single thumb; d, duplex type; d', reduced duplex; t', reduced triplex.

Houdan crosses. For 27 per cent of the latter are non-polydactyl, while, taking the table as it stands, at most only about 4 per cent and (as just stated) probably none of the Silkie offspring were of the typical single-thumbed type. Also the average degree of polydactylism is much greater in the Silkie than in the Houdan crosses. This excess is in part due to the different method of counting toes in the Silkie and the Houdan hybrids; for whereas in the latter the visible toes are counted as equivalent units, in the former in the case of each reduced type one unit more is assigned than appears. The actual number of toes occurring in the Silkie hybrids was also calculated, and it was found that this still averaged higher than that of the Houdans (9.45 as opposed to 9.26).

In hybrids of both classes the greatest number of toes occurring on one foot never exceeds the greatest number possessed by its parents; indeed, the most polydactyl hybrids of the F_1 generation of Silkies never have as many as 6 toes on one foot. This result is not to be explained as due to a regression towards the 4–4-toed condition, but rather as due to the intermediate condition of the heterozygote. For 80 per cent of the hybrids show either the typical or the reduced D type on one or both feet, although neither parent exhibits these types.

We have next to consider the results of mating together the F_1 hybrids. Table 11 gives the results of all matings of this sort.

TABLE 11.—*Frequency of the various types of toes in the second hybrid generation between normal and extra-toed races. Lettering as in table 10.*

A. HOUDAN CROSSES ($F_1 \times F_1$).

Serial No.	Pen No.	Mother			Father			Offspring					Average num. of toes per bird.
		No.	Races involved.	No. of toes.	No.	Races involved.	No. of toes.	Types of toes.					
								4–4	4–5	5–5	4–6	5–6	
1	631	429	Houd.×Wh. Legh..	5–5	83	Wh. Legh.×Houd..	4–4	*14	7	28	1	...	9.3
2	728	174	....Do.........	5–5	258	....Do........	5–5	11	1	20	...	...	9.3
3	631	448	....Do.........	5–5	409	....Do........	4–4	13	4	18	...	...	9.1
4	637	529	Houd.×Min.....	5–5	570	Houd.×Min.....	4–4	4	..	5	...	...	9.1
5	631	430	Houd.×Wh. Legh..	4–4	83	Wh. Legh.×Houd..	4–4	20	1	21	...	...	9.0
6	631	504	Wh. Legh.×Houd..	5–5	83	....Do........	4–4	27	3	23	...	...	8.9
7	631	174	Houd.×Wh. Legh..	5–5	83	....Do........	4–4	14	9	11	...	1	8.9
8	519	85	....Do.........	4–5	83	....Do........	4–4	9	2	4	...	...	8.7
9	637	569	Houd.×Min.....	5–5	570	Houd.×Min.....	4–4	14	1	4	...	1	8.7
10	637	797	....Do.........	5–5	570	....Do........	4–4	2	..	1	...	...	8.7
11	631	86	Houd.×Wh. Legh..	4–4	83	Houd.×Wh. Legh..	4–4	11	1	6	...	...	8.7
12	637	685	Houd.×Min.....	4–4	570	Houd.×Min.....	4–4	5	1	2	...	...	8.6
13	631	84	Houd.×Wh. Legh..	4–4	83	Houd.×Wh. Legh..	4–4	17	13	4	...	...	8.6
14	519	84	....Do.........	4–4	83	....Do........	4–4	7	1	2	...	...	8.5
15	519	86	Wh. Legh.×Houd..	4–4	83	Wh. Legh.×Houd..	4–4	12	2	2	...	...	8.4
			Totals (380)................					180	46	151	1	2	8.92
			Percentages................					47.4	12.1	39.7	0.3	0.5	.....

B. SILKIE CROSSES ($F_1 \times F_1$).

Serial No.	Pen No.	Mother			Father			Offspring (F_2).										
		No.	Races.	No. of toes.	No.	Races.	No. of toes.	Types of toes.										
								ss.	sd.	d'd'	d'd.	dd.	st.	d't'.	dt'.	dt.	t't'.	t't.	tt.
16	753	2071	Min.×Silk....	4–4	2573	Min.×Silk....	4–5	7	..	..	1	19	..	1	..	3	..	1	..
17	753	1966	....Do.......	4–4	2573	....Do.......	4–5	12	2	..	..	15	1	..	..	2	..	..	4
18	753	2575	....Do.......	4–5	2573	....Do.......	4–5	18	..	1	..	16	..	..	1	..	..	..	1
19	709	3827	Silk.×Span...	4–4	1578	Silk.×Span...	6–6	3	..	..	..	2	..	..	..	..	..	..	..
20	709	1963	....Do.......	4–4	1578	....Do.......	6–6	12	5	..	..	15	1	..	..	1	..	..	1
21	821	7413	Silk.×Coch...	5–5	6095	Silk.×Coch...	5–5	1	..	..	1	7	..	..	..	2	..	..	1
22	821	7423	....Do.......	5–5	6095	....Do.......	5–5	3	..	..	..	7	..	..	..	..	1	..	1
23	821	7428	....Do.......	5–5	6095	....Do.......	5–5	5	..	1	..	4	13	..	..	2	..	..	1
24	821	7406	....Do.......	5–5	6095	....Do.......	5–5	3	1	..	..	8	..	..	..	1	1	..	..
			Total (206)................					64	8	2	7	102	2	1	3	8	2	1	8

* Includes 1 case of 3–4 toes.

Comparing tables 10 and 11, it is at once clear that in the second hybrid generation the proportion of extra-toed offspring has decreased. This accords with expectation, if extra-toe is dominant, for then only 75 per cent would be of the dominant type in F_2, while 100 per cent would be of that type in F_1.

Table 12 will enable us to analyze the difference of the proportions in tables 10 and 11.

TABLE 12.—*Percentages of the various types of toes in F_1 and F_2 of the polydactyl hybrids compared.*

No. of toes.	a. Houdan hybrids.		b. Silkie hybrids (as observed).		c. Silkie hybrids (as interpreted).*	
	F_1.	F_2.	F_1.	F_2.	F_1.	F_2.
4–4	27.3	47.4	9.4	31.7	4.3	30.8
4–5	19.1	12.1	9.4	7.7	2.9	3.8
4–6	....	.3	....	1.0	1.5	1.0
5–5	53.6	39.7	81.2	51.4	76.8	53.4
5–6	....	.5	....	4.3	5.8	5.8
6–6	....	....	....	3.9	8.7	5.3

* Reduced duplex and triplex toes classified as typical duplex and triplex.

These tables yield several points of interest. First, although the proportions of normal and extra toe in table 12, *a* and *c*, are not Mendelian, yet the average *increase*, from F_1 to F_2, in the proportion of the recessive (4-toed) type is almost exactly what is called for by Mendel's law. That law calls for an increase of 25 per cent. The actual average increase is 23.3 per cent (20.1 and 26.5 in the two cases). It seems fair to conclude, consequently, that Mendel's law does hold here, and that the 4-toed individuals of F_1 are heterozygotes with imperfect dominance. The feet of most of the 4-toed Silkies of this generation belong, indeed, to the reduced 5-toed type (table 10, B), and the reduced condition is *prima facie* evidence of heterozygotism. In F_1 Silkies of the first hybrid generation, 20 per cent of the feet exhibit "reduced" types of toes, but in F_2 only 5 per cent; and this might have been anticipated, since in F_2 heterozygotes are relatively only half as numerous as in F_1. Again, in F_2 we see reappearing the high ancestral toe-numbers (practically lost in the heterozygotes of F_1, table 12, *b*). These I interpret as extracted dominants. 6-toed extracts are more numerous among the Silkie than the Houdan hybrids, because the Silkie ancestors were 6-toed and the Houdan ancestors only 5-toed. However, only a small proportion of the extracted Silkie dominants have as many toes as the original Silkie ancestors, and this indicates a permanent regression (through the contaminating influence of hybridization?) toward the normal condition of toes. It will be observed that, although 6 toes are not found in the Silkie hybrids of F_1, many of these heterozygotes are of the reduced triplex type. Classifying them as virtually 6-toed, we find (table 12, *c*) 14.5 per cent of the 6-toed type in the F_1 generation.

Among the extracted dominants of F_2 are a few showing more toes than appeared in the ancestors (table 12, *a*; there was also one 7-toed F_2 Silkie hybrid, not recorded in the table). It is this sort of an advance in F_2 that permits the breeder to make a forward step. Theoretically, the appearance of this more aberrant class is probably due to the greater numbers of progeny than of ancestors, since the extracted dominants of

TABLE 13.—*Distribution of toe-numbers in the offspring of DR × R matings.*

A. HOUDAN CROSSES.

Serial No.	No. of pen.	Mother.			Father.			Offspring.				
		No.	Races involved.	No. of toes.	No.	Races involved.	No. of toes.	4–4 toes.	4–5 toes.	5–5 toes.	4–6 toes.	Average num. of toes per bird.
1	519A	87	Houd.×Wh. Legh...	4–5	71	Wh. Legh.........	4–4	17	2	6	...	8.6
2	671	742	Min.×Dk. Brah....	4–4	352	Houd.×Dk. Brah...	4–4	8	2	2	...	8.5
			Totals (37)............................					25	4	8	...	8.54

B. SILKIE CROSSES.

3	706	10	Wh. Legh..........	4–4	1965	Silkie×Spanish......	5–5	4	...	4	...	9.00
4	766	3814	....Do.............	4–4	834	Blk. Game×Silkie...	5–5	10	4	8	1	9.00
5	766	10	....Do.............	4–4	834	....Do..............	5–5	7	...	5	...	8.83
6	607	203	Frizzle×Silkie......	5–5	15	Frizzle.............	4–4	15	2	9	...	8.77
7	766	3815	Wh. Legh..........	4–4	834	Blk. Game×Silkie...	5–5	11	...	7	...	8.77
8	706	3815	....Do.............	4–4	1965	Silkie×Spanish......	5–5	6	...	3	...	8.67
9	706	71	....Do.............	4–4	3823	....Do..............	5–5	18	1	8	...	8.63
10	766	3832	Buff Legh..........	4–4	834	Blk. Game×Silkie...	5–5	7	...	2	...	8.44
11	706	3833	....Do.............	4–4	1965	Silkie×Spanish......	5–5	3	1	...	...	8.25
12	607	230	Frizzle×Silkie......	4–4	15	Frizzle.............	4–4	23	2	2	...	8.22
13	706	71	Wh. Legh..........	4–4	1965	Silkie×Spanish......	5–5	5	...	...	...	8.00
14	706	3814	....Do.............	4–4	1965	....Do..............	5–5	6	...	...	...	8.00
15	706	3832	Buff Legh..........	4–4	1965	....Do..............	5–5	5	...	...	...	8.00
			Totals (179)..........................					120	10	48	1	8.60

TABLE 14.—*Distribution of toe-numbers in the offspring of DR × D matings.*

A. HOUDAN CROSSES.

Serial No.	Pen No.	Mother.			Father.			Offspring.					
		No.	Races involved.	No. of toes.	No.	Races involved.	No. of toes.	4–4 toes.	4–5 toes.	5–5 toes.	5–6 toes.	6–6 toes.	Average num. of toes per bird.
1	803	529	Houdan×Min....	5–5	7522	Houdan..........	5–5	1	4	13	...	...	9.67

B. SILKIE CROSSES.

2	606	182	Frizzle×Silkie....	4–4	775	Silkie...........	6–6	...	3	10	3	5	10.48
3	606	182	....Do............	4–4	21A	....Do...........	6–6	...	...	5	...	1	10.33
4	606	182	....Do............	4–4	551	....Do...........	5–6	...	...	5	...	...	10.00
			Totals (32)............................					...	3	20	3	6	10.36

TABLE 15.—*Percentages of the various types of toes in F_1, F_2, DR × R, and DR × D matings of the polydactyl crosses compared.*

No. of toes.	a. Houdan crosses.				b. Silkie crosses.				c. Silkie crosses (reduced forms of toe classified as typical).			
	Mating F_1	Mating F_2	Mating DR×R	Mating DR×D	Mating F_1	Mating F_2	Mating DR×R	Mating DR×D	Mating F_1	Mating F_2	Mating DR×R	Mating DR×D
	P. ct.	P. ct.	P. ct.	P. ct.	P. ct.	P. ct.	P. ct.	P. ct.	P. ct.	P. ct.	P. ct.	P. ct.
4–4	27.3	47.4	67.6	5.6	9.4	31.7	67.0	....	4.3	30.8	66.7	....
4–5	19.1	12.1	10.8	22.2	9.4	7.7	5.6	9.4	2.9	3.8	3.1	9.4
5–5	53.6	39.7	21.6	72.2	81.2	51.4	26.8	62.5	76.8	53.4	24.6	62.5
4–6	....	.3	....	....	....	1.0	.6	....	....	1.0	1.9	....
5–6	....	.5	....	....	....	4.3	....	9.4	5.8	5.8	1.5	9.4
6–6	....	....	....	....	....	3.9	....	18.7	8.7	5.3	1.2	18.7
6–7	....	....	....	....	....	....	....	....	....	....	....	....

F_2 are seven times as numerous as their extra-toed grandparents. Here, as elsewhere, the absolute range of variability depends upon the number of individuals observed.

As we have seen, failure of dominance is much more complete in some of the individuals of F_2, namely, those with 4 toes, than others. There is a variation in "potency." Is the degree of potency inherited? Do the 4-toed heterozygotes produce a larger proportion of imperfect dominants in F_3 than the 5-toed heterozygotes? The answer to this question should be given by the correlation between total number of toes in the two parents and average number of toes in their offspring, as given in table 11. In the case of the Houdan crosses there is a strong positive correlation, measured by 0.683 ± 0.092; but the correlation is insignificant in the Silkie crosses (-0.085 ± 0.032). This lack of correlation in the Silkie hybrids is perhaps due to the heavy regression in toe-number characteristic of the second hybrid generation. In general, there seems to be an inheritance of potency.

It now remains to test our conclusions by reference to the mating of the heterozygote with the dominant and with the recessive types, respectively. An examination of tables 13 to 15, particularly the last, reveals several points of interest. Mendelian expectation in the DR × R cross is 50 per cent of the recessive (4–4) type. Actually, in the two crosses, A and B, 68 per cent and 67 per cent, respectively, were obtained. But recalling that of these amounts one-half of 27.3, or 13.71, and one-half of 9.4, or 4.7, are respectively due to failure to develop the extra-toe in heterozygotes, there remain 54 per cent and 62 per cent, respectively, of 4-toed offspring, which doubtless represent the extracted RR type and approach the expected proportions.

Mendelian expectation in the DR × D cross (table 15) is 50 per cent heterozygotes and 50 per cent extracted dominants. Of the heterozygotes some 14 per cent may be expected to show 4–4 toes; that the percentage is much less than that is doubtless due to the small numbers involved. What is striking is the reappearance, in the second generation, of large proportions of the extreme dominant type. These results thus confirm those of the F_2 generation.

Since extra-toe frequently fails to dominate, there should be certain 4-toed heterozygotes which throw extra-toe offspring, and such are found. In table 16 are given six matings of 4-toed DR's. One sees that they produce some 5-toed offspring. On the other hand, extracted 4-toed recessives are obtained, as table 17 shows.

Finally, we must consider whether, among the polydactyl birds of one class, *e. g.*, Houdans or Silkies, there is any difference in the "centgener power" of parents corresponding to the degree of development of their extra toes. This inquiry is suggested by Castle's study (1906, p. 20) of polydactyl guinea-pigs. He finds that when the extra toes of the mothers are graded into the 5 classes, good (G), fair (F), poor (P), normal though of abnormal ancestry (N), and normal of normal ancestry (N'), it follows: "first, that the proportion of polydactylous young produced by a male decreases in the successive classes from G to N'; and, secondly, that the degree of development of the toes produced on those polydactylous young diminishes in the same order." It is possible to test this conclusion in poultry because, inside of any one type of extra-toe, *e. g.*, the triplex type, variation appears in the absolute size of the toes and in the degree of their separateness. Our questions, then, are: (1) does the *proportion of polydactyl young* produced by a pair of birds of any type diminish with the degree of development of toes inside of that type, and (2) does the *degree of development* of the toes produced on the polydactylous offspring diminish in the same order?

TABLE 16.—*Distribution of toe-numbers in the offspring of 4-toed heterozygotes.*

Pen No.	Mother.			Father.			Offspring.			Nature of mating.
	No.	Races.	No. of toes.	No.	Races.	No. of toes.	4-4 toes.	4-5 toes.	5-5 toes.	
637	685	Houd.×Min.	4-4	570	Houd.×Min.	4-4	5	1	2	DR×DR
729	913	Houd.×Min.	4-4	936	Houd.×Legh.	4-4	38	13	19	DR×DR
729	2269	....Do..........	4-4	936	....Do..........	4-4	15	5	10	DR×DR
729	2324	....Do..........	4-4	936	....Do..........	4-4	30	5	3	DR×R
642	750	Min.×Polish	4-4	647	....Do..........	4-4	10	...	3	R×DR
671	742	Min.×Brah.	4-4	352	Houd.×Brah.	4-4	8	2	2	R×DR

TABLE 17.—*Distribution of toe-numbers in the offspring of extracted 4-toed parents.*

Pen No.	Mother.			Father.			Offspring.			Nature of mating.
	No.	Races.	No. of toes.	No.	Races.	No. of toes.	4-4 toes.	4-5 toes.	5-5 toes.	
762	2011	Polish×Min.	4-4	444	F₁ Houd.×Legh.	4-4	10	...	...	R×R
	2614	....Do..........	4-4	444	....Do..........	4-4	6	...	...	R×R
	2333	....Do..........	4-4	444	....Do..........	4-4	16	...	...	R×R
	2618	....Do..........	4-4	444	....Do..........	4-4	2	...	...	R×R
	3776	....Do..........	4-4	444	....Do..........	4-4	2	...	...	R×R

Two sets of data are available for answering these questions. The most direct set includes the data derived from crossing "pure-bred" polydactyl birds and the other includes the data derived from using hybrids between normal-toed and polydactyl ancestors. The latter data have the advantage that the parents offer a greater variability; but they have the disadvantage that the germinal condition of those parents is incompletely known.

The pure races may be considered first. Eight matings of Houdans, each parent with 5 toes, gave 122 offspring, of which 116 had 5-5 toes, 3 had 4-5 toes, and 3 had 4-4 toes. The variability of the toes is not great in the parent Houdans. But, arranging them in the order of development of the toes, the most developed first, the series of table 18 results.

TABLE 18.

Serial No.	Pen No.		No. of mother.	Offspring.			
				4-4 toes.	4-5 toes.	5-5 toes.	Average.
1	727	803	2457	1	2	34	9.89
2	727	803	3105	1	0	45	9.95
3	803		2579	..	1	12	9.92
4	727		3106	..	..	4	10.00
5	727		2494	1	0	5	9.67
6	727		2459	..	..	16	10.00

No direct relation here appears between development of the extra toe in the parents and the average number of toes in the offspring.

Of the Silkies, 3 hens were used in 5 matings. The same 6-toed cock (No. 774) was employed throughout (table 19).

TABLE 19.

Serial No.	Pen No.		Mother.		f		Offspring.					
			No.	No. of toes.			4-4 toes.	5-4 toes.	5-5 toes.	4-6 toes.	5-6 toes.	6-6 toes.	Average.
1	734	815	499	6-6	21	a	2	1	7	0	3	8	10.3
						b	1	0	3	0	0	17	11.4
2	734	815	773	6-5	13	a	..	..	6	0	3	4	10.9
						b	..	..	2	1	1	9	11.4
3	734		500	5-5	8	a	..	2	4	0	2	..	10.0
						b	..	..	3	2	2	1	10.5

In table 19 the series a of observed average numbers of filial toes (10.3, 10.9, 10.0) and the series b obtained by assigning the typical full number to all reduced types (11.4, 11.4, 10.5) are decidedly irregular. There is, however, between the parental and the filial series a correlation of $+0.250 \pm 0.070$. This indicates a slight tendency for the number of toes in the progeny to vary with those of the parentage.

The second set of data is derived from special matings made with hybrids between Houdans and 4-toed races. On the one hand, in pens 728 and 813, cocks with well-developed toes of the duplex type were mated with hens as nearly as possible of the same sort; while in pens 765, 769, and 820 cocks with small, imperfectly separated toes (probably of the duplex type *) were mated with hens as far as possible of the same sort.

Tables 20, 21, and 22 give in detail and in summary the distribution of types of polydactylism in the families from well-developed and in those from poorly developed parents. They show a great difference between the offspring of parents with good extra-toe (table 20) and those with poor extra-toe (table 21). The former yield over 80 per cent offspring with 5 toes or more on one or both feet, while the latter yield about 57 per cent of such.

---

* I say probably of the duplex type because the cock of pen 769 had a distally split toe on the right foot, reminding somewhat of the reduced triplex type. But as the left foot had a typical duplex thumb, and the triplex is not common in Houdans, it should probably be classed as duplex.

On the other hand, in the former families there are less than half as many offspring with only 4 toes as in the latter. Classifying "reduced" forms with their proper advanced type, we find highly polydactyl parents yielding only 16 per cent non-polydactyl offspring, while slightly polydactyl parents yield 43 per cent non-polydactyl offspring. The percentage of polydactylous young diminishes with the size and distinctness of the extra toes and the grades of the polydactyl offspring are lower (absence in table 22, *b*, of 6 toes). Both of Castle's conclusions seem to be confirmed.

TABLE 20.—*Distribution of toe-types in the offspring of "good" extra-toed parents.*

Serial No.	Pen No.	Mother No.	Mother Gen.	Mother Races	Father No.	Father Gen.	Father Races	Mating	4–4	4–5	5–5	5–6	6–6	Average	ss.	sd.	d'd'	d'd.	dd.	t't'.	dt'.	dt.	t't'.	tt.	q't
1	728	2271	F₂	Wh.Legh.×Houd.	258	F₁	Houd.×Wh.Legh.	DD×DR	4	1	21	..	..	9.65	3	..	1	1	21	..	..	..	..	..	..
2	728	912	F₂	Do.	258	F₁	Do.	DR×DR	5	3	21	..	..	9.55	5	3	..	..	20	1	..	..	..	..	..
3	728	2248	F₂	Do.	258	F₁	Do.	DD×DR	8	3	22	..	..	9.42	8	3	..	..	21	..	..	..	1	..	..
4	728	2272	F₂	Do.	258	F₁	Do.	DR×DR	17	4	34	..	..	9.31	17	1	..	3	34	..	..	..	..	..	..
5	728	174	F₁	Do.	258	F₁	Do.	DR×DR	10	1	15	..	..	9.19	10	1	..	..	14	1	..	..	..	..	..
				Totals (169)					44	12	113	..	..	9.41	43	8	1	4	110	0	2	0	1	..	..
				Percentages					26.0	7.1	66.9	..	..		25.4	4.7	0.6	2.4	65.2	..	1.2	..	0.6	..	..
6	813	2271	F₂	Wh.Legh.×Houd.	3904	F₂	Houd.×Wh.Legh.	D×D	..	2	32	..	..	9.94	..	..	..	2	32	..	..	..	..	..	..
7	813	5113	F₂	Do.	3904	F₂	Do.	D×D	2	1	32	1	..	9.89	..	2	1	32	..	1	..	..	..	..	..
8	813	377	F₁	Do.	3904	F₂	Do.	DR×D	2	5	17	..	1	9.68	2	2	..	3	16	..	1	..	..	1	..
9	813	5122	F₂	Do.	3904	F₂	Do.	D×D	1	3	7	..	..	9.55	1	3	..	..	7	..	..	..	..	..	..
10	813	935	F₂	Do.	3904	F₂	Do.	DR×D	1	2	25	1	1	9.53	1	2	..	..	25	..	1	..	..	..	1
11	813	2272	F₂	Do.	3904	F₂	Do.	DR×D	5	2	18	..	..	9.52	4	1	..	..	18	1	..	..	..	..	..
12	813	912	F₂	Do.	3904	F₂	Do.	DR×D	4	5	11	..	..	9.35	3	5	1	..	11	..	..	..	..	..	..
13	813	7320	F₂	Do.	3904	F₂	Do.	DR×D	5	1	11	..	..	9.35	3	1	2	..	11	..	..	..	..	..	..
14	813	5142	F₂	Do.	3904	F₂	Do.	DR×D	2	1	4	..	..	9.28	2	..	..	1	4	..	..	..	..	..	..
				Totals (205)					22	22	157	2	2	9.70	16	14	6	7	156	1	1	2	0	1	1
				Percentages					10.7	10.7	76.5	1.0	1.0		7.8	6.8	2.9	3.4	76.2	0.5	0.5	1.0	..	0.5	0.5

TABLE 21.—*Distribution of toe-types in the offspring of "poor" extra-toed parents.*

Serial No.	Pen No.	Mother No.	Mother Gen.	Mother Races	Father No.	Father Gen.	Father Races	Mating	4–4	4–5	5–5	5–6	Average	ss.	sd.	d'd'	d'd.	dd.	t't'.	dq'.
1	765	984	F₁	Wh.Legh.×Houd.	1794	F₂	Wh.Legh.×Houd.	DR×DR	9	5	11	..	9.08	9	3	..	2	10	1	..
2	765	1790	F₂	Do.	1794	F₂	Do.	DR×DR	18	7	17	..	8.98	18	6	..	1	17	..	..
				Totals (67)					27	12	28	..	9.02	27	9	..	3	27	1	..
				Percentages					40.3	17.9	41.8	..		40.3	13.4	..	4.5	40.3	1.5	..
3	769	492	F₁	Wh.Legh.×Houd.	911	F₂	Wh.Legh.×Houd.	DR×DR	13	1	14	..	9.04	13	1	..	..	14	..	..
4	769	4976	F₂	Do.	911	F₂	Do.	DR×DR	11	3	9	..	8.91	11	3	..	..	8	1	..
5	769	2254	F₂	Do.	911	F₂	Do.	DR×DR	22	6	8	..	8.61	22	4	..	2	8	..	..
6	769	1305	F₂	Do.	911	F₂	Do.	DR×DR	12	1	4	..	8.53	12	..	..	1	4	..	..
				Totals (104)					58	11	35	..	8.77	58	8	..	3	34	1	..
				Percentages					55.8	10.6	33.7	..		55.8	7.7	..	2.9	32.7	1.0	..
7	820	984	F₁	Wh.Legh.×Houd.	4731	F₂	Wh.Legh.×Houd.	D×DR	2	3	27	..	9.78	2	2	..	1	27	..	..
8	820	2255	F₂	Do.	4731	F₂	Do.	DR×DR	6	1	10	..	9.24	6	..	..	1	10	..	..
9	820	6479	F₂	Do.	4731	F₂	Do.	DR×DR	12	2	16	..	9.13	10	1	2	1	15	1	..
10	820	2016	F₁*	Do.	4731	F₂	Do.	DR×DR	9	2	2	..	8.45	9	2	..	..	2	..	..
				Totals (92)					29	8	55	..	9.28	27	5	2	3	54	1	..
				Percentages					31.5	8.7	59.8	..		29.3	5.4	2.2	3.3	58.7	1.1	..

* No. 2016 has 4–4 toes and is a hybrid between a 5-toed White Leghorn × Houdan and a 4-toed Minorca × Polish.

But a more critical examination of the parentages of the 5 pens shows that they are not comparable. In matings 6 to 14 of table 20 the cock is almost certainly a dominant in respect to toes; whereas the cocks in table 21 are probably heterozygous. The heterogygous state determines two things: the imperfect nature of the extra-toe and a relative deficiency in the offspring of the higher toe-numbers. In our results we can not say that one of these things is the cause of the other, as Castle does; they are, rather, in all probability, due to a common cause. I think Castle's paper may justly be criticized for not giving sufficient data concerning the ancestry of the individual mothers used. Without such data the paper can not be said satisfactorily to demonstrate his conclusion.

TABLE 22.—*Summary of observed toe-numbers in offspring, percentages.*

a. Parents have "good" extra toes.						b. Parents have "poor" extra toes.			
Pen No.	4-4 toes.	4-5 toes.	5-5 toes.	5-6 toes.	6-6 toes.	Pen No.	4-4 toes.	4-5 toes.	5-5 toes.
728	26.0	7.1	66.9	....	....	765	40.3	17.9	41.8
813	10.7	10.7	76.5	1.0	1.0	769	55.8	10.6	33.7
						820	31.5	8.7	59.8
Average..	17.7	9.1	72.2	0.5	0.5	Average..	43.2	11.8	44.9

To summarize: "Potency," as measured by dominance of the extra-toed condition, is inherited, in the Houdan crosses at least. There is some evidence, derived from "pure-bred" Silkies, that differences in the degree of development of the extra-toes are inherited. But the average condition of the toes in the offspring of second or later generation hybrids can not be used as evidence of inheritance of the degree of parental development of the toes, since these are dependent on the same basal cause, namely, the hidden gametic constitution of the parents. Despite the obscuration of imperfect dominance, polydactylism in poultry proves itself to be a unit-character that segregates.

CHAPTER III.

SYNDACTYLISM.

A. STATEMENT OF PROBLEM.

In man, various mammals, and some birds two or more adjacent fingers are sometimes intimately connected by an extension of the web that is normally a mere rudiment at their base. Such a condition is known as syndactylism. A good introductory account of syndactylism is given by Bateson (1904, pp. 356-358). Taking a number of cases of syndactylism together, he says: "A progressive series may be arranged showing every condition, beginning from an imperfect webbing together of the proximal phalanges to the state in which two digits are intimately united even in their bones, and perhaps even to the condition in which two digits are represented by a single digit." He also calls attention to the fact that in the human hand "there is a considerable preponderance of cases of union between the digits III and IV;" while in the foot the united digits "are nearly always II and III." The matter of syndactylism in birds has a peculiar interest because of the fact that among wading and swimming birds syndactylism has become a normal condition of the feet, and, moreover, just this feature is one that has become classical in evolutionary history, because Lamarck thought it well illustrated his idea of the origin of an organ by effort and use.

Concerning the cause of syndactylism little can be said. Both in mammals and birds the digits are indicated before they are freed from lateral tissue connections. The linear development of the fingers is in part accompanied by a cutting back of this primordial web, in part by a growth beyond it. In syndactylism growth of the web keeps pace with that of the fingers. From this point of view syndactylism may be regarded as due to a peculiar excessive development of the web.* In some human cases adhesions of the apex of the appendage to the embryonic membranes has stimulated the growth of the interdigital membrane, resulting in syndactylism. But it would be absurd to attempt to explain syndactylism in general on this ground. The more "normal" forms of syndactylism, as seen in poultry, still want for a causal explanation.

Most of the cases of syndactylism whose inheritance is about to be described arose in a single strain of fowl and can, indeed, be traced back to a single bird. This ancestor is No. 121, a Dark Brahma hen described in a previous report.† It was only in the search for the origin of the exaggerated forms of syndactylism observed in some of her descendants that an unusu-

---

*Lewis and Embleton (1908, p. 45) present strong arguments against the theory that syndactylism is due to arrested development.

† Davenport, 1906, page 34, Plate V

TABLE 23.—*Ancestry of syndactyl fowl and the results of various matings involving syndactylism.*

[Abbreviations: Aba, Abβ, etc., types of syndactylism (p. 32); F, father; FF, father's father; FM, father's mother; M, mother; MF, mother's father; MM, mother's mother; M×P, hybrid of Minorca and Polish races; Synd., syndactyl (type unknown). f, foot. In Nos. 24 to 42 two cocks (Nos. 242 and 3116, and 5399 and 4562, respectively) were at different times used.]

Serial No.	Pen No.	First mating.									Second mating.									Average per cent syndactyl.
		Ancestry.						Offspring.			Ancestry.						Offspring.			
		M's No.	M.M.	MF.	F's No.	FM.	FF.	Syndactyl.			M's No.	MM.	MF.	F's No.	FM.	FF.	Syndactyl.			
								2f.	1f.	0f.							2f.	1f.	0f.	
1a, b	627	302	[1]121	[2]8A	180	[1]121	[2]8A	0	0	34	302	[1]121	[2]8A	242	[1]121	[2]8A	3	0	29	10.3
2a, b	627	280	121	8A	180	121	8A	0	0	23	280	121	8A	242	121	8A	2	0	21	9.5
3a, b	627	181	121	8A	180	121	8A	0	0	20	181	121	8A	242	121	8A	3	0	33	9.1
4a, b	627	354	121	8A	180	121	8A	0	0	24	354	121	8A	242	121	8A	1	0	37	2.6
5a, b	627	178	121	8A	180	121	8A	0	0	20	178	121	8A	242	121	8A	0	0	42	....
6a, b	627	190	121	8A	180	121	8A	1	0	24	190	121	8A	242	121	8A	0	0	6	....
7a, b	...	353	121	8A	180	121	8A	0	0	13	353	121	8A	242	121	8A	0	0	22	....
8a, b	...	300	121	1A	180	121	8A	0	0	23	300	121	1A	242	121	8A	0	0	37	....
Totals (182)							1	0	181	Totals (236)						9	0	227	
Percentages							0.55	...	99.45	Percentages						3.81	...	96.19	

Serial No.	Pen No.	Mother.			Father.			Offspring.								
		No.	Bred in pen No.	Toes.	No.	Bred in pen No.	Toes.	Syndactyl.				Classification.				
								2f.	1f.	0f.	P. ct.	Aaa.	Aba.	Abβ.	Abβ'.	Bba.
9	747	2526	[3]658	Normal.	1888	[3]658	Normal.	9	0	9	50.0	....	2	16	....	....
10	747	2831	658	...Do...	1888	658	...Do...	6	0	6	50.0	....	7	5	....	....
11	747	2652	658	...Do...	1888	658	...Do...	3	0	25	10.7	....	6	....	....	....
12	747	3541	658	...Do...	1888	658	...Do...	4	0	41	8.9	1	4	3	....	....
13	747	1892	658	...Do...	1888	658	...Do...	4	0	47	7.8	....	....	....	....	....
14	747	1872	658	...Do...	1888	658	...Do...	0	0	28	0.0	....	....	....	....	....
15	747	1874	658	...Do...	1888	658	...Do	0	0	28	0.0	....	....	....	....	....
								26	0	184	12.4					
16	703	2353	D. Br.	...Do...	122	D. Br.	...Do	1	0	6	14.3	....	2	....	....	....
17	703	2030	D. Br.	...Do...	122	D. Br.	...Do	2	1	12	20.0	....	5	....	....	....
								3	1	18	18.2					
18	754	3126	[4]627	Normal.	871	[4]627	Normal.	12	1	30	30.2	....	13	12	....	....
19	754	3175	627	...Do...	871	627	...Do...	3	0	8	27.3	....	3	3	....	....
20	754	873	627	...Do...	871	627	...Do...	[2]	(?)	(?)	(?)	....	....	4	....	....
21	754	1052	627	...Do...	871	627	...Do...	0	0	17	0.0	....	....	....	....	....
22	754	853	627	...Do...	871	627	...Do...	0	0	19	0.0	....	....	....	....	....
23	754	862	627	...Do...	871	627	...Do...	0	0	27	0.0	....	....	....	....	....
								15	1	101	13.7					
24	767	2526	[3]658	Normal.	3116	D. Br.	Synd....	5	0	22	18.5	1	1	6	....	2
25	767	872	[4]627	Abβ.....	242	[5]513	Normal.	1	0	1	50.0	....	1	....	1	....
25a	767	872	627	Abβ.....	3116	D. Br.	Synd....	7	1	30	21.0	3	5	3	....	4
26	767	2104	[7]608	Normal.	3116	D. Br.	...Do...	3	0	18	14.3	....	2	2	....	2
27	767	2831	[3]658	...Do...	3116	D. Br.	...Do...	3	0	32	8.6	....	6	....	....	....
28	767	181	[5]513	...Do...	242	513	Normal.	1	0	22	4.4	2	....	....	....	....
28a	767	181	513	...Do...	3116	D. Br.	Synd....	1	1	60	3.2	....	1	1	....	1
29	767	190	[5]520	...Do...	242	513	Normal.	1	1	28	6.7	1	....	....	....	2
29a	767	190	520	...Do...	3116	D. Br.	Synd....	4	..	49	7.6	....	3	4	....	1
Syndactyl (242 ♂)								3	1	51	7.3					
Syndactyl (3116 ♂)								23	2	211	9.4					

[1] No. 121 is a Dark Brahma.
[2] No. 8A is a Tosa fowl (Game).
[3] (White Leghorn × Rose Comb Black Minorca) × Dark Brahma.
[4] Dark Brahma.
[5] See *supra*.
[6] 121 ♂ Dark Brahma × 8A Tosa.
[7] F₁ (White Leghorn × Dark Brahma).

TABLE 23.—*Ancestry of syndactyl fowl and the results of various matings involving syndactylism*—Continued.

Serial No.	Pen No.	Mother.			Father.			Offspring.								
								Syndactyl.				Classification.				
		No.	Bred in pen No.	Toes.	No.	Bred in pen No.	Toes.	2f.	1f.	0f.	P. ct.	Aaa.	Aba.	Abβ.	Abβ'.	Bba.
30	801	4569	767	Aba	5399	747	Aba	2	0	0	100.0	1	0	3	0	0
30a	801	4569	767	Aba	4562	767	Normal	0	2	2	50.0	....	1	1	....	....
31	801	6843	767	Normal	4562	767	...Do	1	3	2	66.7	....	2	2	1	....
32	801	872	627	Abβ	5399	747	Aba	12	4	11	59.3	3	9	11	....	5
32a	801	872	627	Abβ	4562	767	Normal	7	1	12	40.0	2	8	4	1	....
33	801	5515	767	Bba	5399	747	Aba	4	0	7	36.4	....	2	6	....	....
33a	801	5515	767	Bba	4562	767	Normal	1	2	5	37.5	2	1	1	....	....
34	801	7528	767	Aba	5399	747	Aba	1	0	0	100.0	....	2	....	....	....
34a	801	7528	767	Abβ	4562	767	Normal	2	1	7	30.0	....	1	4	....	....
35	801	6861	767	Normal	4562	767	...Do	1	0	3	25.0	....	2	....	....	....
36	801	6869	767	...Do	5399	747	Aba	0	1	3	25.0	1	....	....	....	....
36a	801	6869	767	...Do	4562	767	Normal	1	0	4	20.0	....	....	2	....	....
37	801	2831	658	...Do	5399	747	Aba	3	1	18	18.2	....	4	....	....	3
37a	801	2831	658	...Do	4562	767	Normal	2	1	11	21.4	....	2	....	....	3
38	801	2526	658	...Do	5399	747	Aba	0	0	5	0.0	....	....	....	....	....
38a	801	2526	658	...Do	4562	767	Normal	1	0	2	33.3	....	1	1	....	....
39	801	4570	767	...Do	5399	747	Aba	0	1	5	16.7	1	....	....	....	....
39a	801	4570	767	...Do	4562	767	Normal	0	2	17	10.5	1	1	....	....	....
40	801	1892	658	...Do	5399	747	Aba	0	0	9	0.0	....	....	....	....	....
40a	801	1892	658	...Do	4562	767	Normal	1	0	3	25.0	....	2	....	....	....
41	801	4263	767	...Do	5399	747	Aba	0	1	4	20.0	....	1	....	....	....
41a	801	4263	767	...Do	4562	767	Normal	0	0	10	0.0	....	....	....	....	....
42	801	6872	767	...Do	4562	767	...Do	0	0	6	0.0	....	....	....	....	....
Syndactyl (5399 ♂)								22	8	62	32.6					
Syndactyl (4562 ♂)								17	12	84	25.7					
43	776	2291	Coch.	Normal	2732	Coch.	Normal	2	0	6	25.0	....	2	....	....	2
44	776	2574	Coch.	...Do	2732	Coch.	...Do	..	1	9	10.0	....	1	....	....	....
45	776	2570	Coch.	...Do	2732	Coch.	...Do	..	1	11	8.3	....	1	....	....	....
46	776	2297	Coch.	...Do	2732	Coch.	...Do	..	1	12	7.7	....	....	....	....	1
47	776	2299	Coch.	...Do	2732	Coch.	...Do	1	0	16	5.9	....	2	....	....	....
48	776	2904	Coch.	...Do	2732	Coch.	...Do	0	0	6	0.0	....	....	....	....	....
49	776	2937	Coch.	...Do	2732	Coch.	...Do	0	0	7	0.0	....	....	....	....	....
50	776	2300	Coch.	...Do	2732	Coch.	...Do	0	0	15	0.0	....	....	....	....	....
51	776	2736	Coch.	...Do	2732	Coch.	...Do	0	0	18	0.0	....	....	....	....	....
								3	3	100	5.7					
52	816	121	D. Br.	Aba	122	D. Br.	Normal	3	1	10	28.6	....	1	....	2	4
52a	816	121	D. Br.	Aba	4912	M×P	...Do	0	0	13	0.0	....	....	....	....	....
53	816	5835	D. Br.	Normal	122	D. Br.	...Do	1	0	6	14.3	....	2	....	....	....
54	816	2353	D. Br.	...Do	122	D. Br.	...Do	0	0	7	0.0	....	....	....	....	....
54a	816	2353	D. Br.	...Do	4912	M×P	...Do	0	0	4	0.0	....	....	....	....	....
Syndactyl (122 ♂)								4	1	23	17.9					
Syndactyl (4912 ♂)								0	0	17	0.0					

ally great extension of the web in her feet was noticed. The syndactyl condition of my birds did not, thus, arise *de novo*, but had its origin antecedent to the beginning of the breeding experiments. In addition to this main strain a slight degree of syndactylism has appeared among some of my Cochin bantams.

The types of syndactylism which have appeared in my flock form a rather extensive series. First, (A) the single web, which, in my specimens, always occupies the interspace between digits III and IV. This is the same interval which is most apt to show the web in syndactylism of the human hand, and, it is suggestive to note, it is this interval that is filled in those wading birds that have the single web only between the toes (*e. g.*, *Cursorius*,

Glareola, Vanellus, Squatarola, Charadrius, Limosa, Machetes, Himantopus); second, there is (*B*) the double web, one-seventh as common, which always occupies the interspaces between the digits II–III and III–IV.

On another basis, the syndactyl feet may be classified as: (*a*) toes adherent, web small in extent, and (*b*) toes distant, web broad. I have found the narrow web only between digits III and IV. It is one-eighth as common as the broad-webbed type. The broad, double web approaches closely to the type found normally in swans, geese, and ducks.

Finally, the syndactyl feet may be classified as: α, straight-toed, or β, curve-toed. Class α is to class β in frequency as 2 : 1. In the typical curve-toed syndactyl foot the web between III and IV is complete to the nails of each; in fact, in extreme cases the nails of the two toes are more or less fused together. From the fused nails the middle toe, being the longer, passes in a curve to the distal end of the metatarsus. The D-shaped interspace between the curved III and straight IV toe is filled with the web. In other cases the nails are merely approximated and the middle toe is slightly curved. In three instances (4 per cent of all) the outer toe (IV) is curved toward the (straight) median toe (class β').

As stated, the polydactyl offspring trace back their ancestry to No. 121; her feet both show the double, broad, straight-toed type (*Bba*). We shall attempt in the following paragraphs to trace the heredity of her type of polydactylism and of the others that have subsequently arisen.

B. RESULTS OF HYBRIDIZATION.

In taking up the results of breeding experiments to test the method of inheritance of syndactylism, it will be best first to give in a table all pens in which the character showed itself, with the frequency of the different types of foot in them (table 23).

The history of the syndactyl strain begins with No. 121 ♀ and in the matings 1 to 8 are given the results of crossing together some of her progeny derived from a normal-toed father. This father was either No. 8A or 1A, both full-blooded Tosa (Japanese Game) fowl and without suspicion in either soma or offspring of syndactyl taint. There is no record of trace of syndactylism in the progeny of 121 × 8A (or 1A); but a slightly developed condition of syndactylism may very well have been overlooked by me in this F_1 generation (as I had never thought of such an abnormality), even as I at first overlooked the syndactylism visible in No. 121. But when these F_1 hybrids were mated together (pen 627, serial Nos. 1 to 8) I got, in the different families, from 10 per cent syndactyl offspring down to none at all.

At first sight the suggestion arises that, if inheritance is at all Mendelian, the normal condition is dominant and that the heterozygotes throw again, in pen 627, the syndactylous condition. If this hypothesis were true it would follow that syndactyls bred together should, sometimes at least, throw, even in large families, 100 per cent syndactyl offspring. But only 2 families, Nos. 30 and 34, have yielded 100 per cent syndactyls, and these contained 2 and 1 offspring, respectively; so they are not significant. On the other hand, there are numerous matings of 2 extracted normal-toed parents that have produced only normal-toed offspring (families Nos. 14, 15, 21, 22, 23, including 119 individuals). Consequently the conclusion is favored that normal-foot is recessive and syndactyl-foot dominant, and this shall be our working hypothesis.

On our hypothesis, No. 121 is probably a heterozygote. Mated with the recessive normal, expectation is 50 per cent heterozygous, showing syndactylism; the remainder normal-toed. But dominance is here, as in polydactylism, very imperfect. For this reason and because it was not looked for, no syndactylism was noted in the first hybrid generation. The offspring prove to be of two sorts, however. No. 180 ♂ is a pure recessive, and in 8 matings with as many different sisters of his he got 184 normal-toed to 1 syndactyl. These same sisters, mated to another brother, No. 242, in some cases gave 9 per cent and 10 per cent syndactyl. No. 242 is, consequently, probably a DR and, mated to DR sisters (which constitute according to expectation about one-half of all) gives some DD's, part of which constitute the 9 to 10 per cent of syndactyls. Of course, 25 per cent DD is to be expected; the difference gives a measure in this instance of the imperfection of dominance in the "extracted" as well as "heterozygous" condition.

Matings 9 to 15 (pen 747) are instructive in comparison with the foregoing case. Both parents are derived from pen 658, which contained as breeders a heterozygous Dark Brahma male (No. 146) and various females of non-booted races far removed from suspicion of syndactylism; expectation being an equal number of DR and RR offspring. In pen 747 No. 1888 ♂ acts like a DR, and so do the hens in matings 9 to 13, while the hens in the other 2 matings are doubtless RR's. The former give 17 per cent syndactyl offspring, the latter none at all (in 56 individuals).

Matings 16 and 17 (pen 703) are between pure-bred Dark Brahmas that are probably DR's. About 22 per cent of their offspring are syndactyl—a rather higher proportion than we have found before. Matings 18 to 19 are between progeny of pen 627. In mating 20 the normals were not recorded. The cock in this pen, No. 871, is probably heterozygous, as are also the first two hens, so that nearly 30 per cent of their progeny are syndactyl. From the other 3 hens no syndactyl offspring were obtained. Evidently the two sets of hens have a very different gametic constitution. The existence of two sorts of families is one of the strong arguments for the segregation of this character.

We next come to the pens (matings Nos. 24 to 42) which were especially mated to study the inheritance of syndactylism. I had now, for the first time, two parents with syndactylic feet.

On account of imperfection of dominance decision as to gametic composition of any parent must largely rest on the make-up of the progeny. Table 24 gives the most reasonable classification of the parentages.

TABLE 24.

DD × DD (SYNDACTYL × SYNDACTYL).

Family No.	Mother's No.	Bred in pen No.	Toes.	Father's No.	Bred in pen No.	Toes.	Syndactyl. 2t.	1t.	0t.	P. ct.
30	4569	767	Abα	5399	747	Abα	2	0	0	100.0
34	7528	767	Abβ	5399	747	Abα	1	0	0	100.0
32	872	627	Abβ	5399	747	Abα	12	4	11	59.3
33	5515	767	Bbα	5399	747	Abα	4	0	7	36.4
Totals							19	4	18	74.2

DD × DR.

Family No.	Mother's No.	Bred in pen No.	Toes.	Father's No.	Bred in pen No.	Toes.	2t.	1t.	0t.	P. ct.
31	6843	767	Normal	4562	767	Normal	1	3	2	66.7
30a	4569	767	Abα	4562	767	Do	0	2	2	50.0
33a	5515	767	Bbα	4562	767	Do	1	2	5	44.4
32a	872	627	Abβ	4562	767	Do	7	1	12	42.9
34a	7528	767	Abβ	4562	767	Do	2	1	7	30.0
36	6869	767	Normal	5399	747	Abα	0	1	3	25.0
25a	872	627	Abβ	3116	D. Br.	Synd	7	1	30	21.1
41	4263	767	Normal	5399	747	Abα	0	1	4	20.0
37	2831	658	Do	5399	747	Abα	3	1	18	18.2
39	4570	658	Do	5399	747	Abα	0	1	5	16.7
40	1892	658	Do	5399	747	Abα	0	0	9	0.0
Totals							21	14	97	26.5

DR × DR.

Family No.	Mother's No.	Bred in pen No.	Toes.	Father's No.	Bred in pen No.	Toes.	2t.	1t.	0t.	P. ct.
38a	2526	658	Normal	4562	767	Normal	1	0	2	33.3
35	6861	767	Do	4562	767	Do	1	0	3	25.0
40a	1892	658	Do	4562	767	Do	1	0	3	25.0
37a	2831	658	Do	4562	767	Do	2	1	11	21.4
36a	6869	767	Do	4562	767	Do	1	0	4	20.0
24	2526	658	Do	3116	D. Br.	Synd	5	0	22	18.5
26	2104	608	Do	3116	Do	Do	3	0	18	14.3
39a	4570	767	Do	4562	767	Do	0	2	17	10.5
27	2831	658	Do	3116	D. Br.	Do	3	0	32	8.6
29a	190	520	Do	3116	D. Br.	Do	4	0	49	7.6
29	767	190	Do	242	513	Do	1	1	28	6.7
28a	181	513	Do	3116	Do	Do	1	1	60	3.2
Totals							23	5	249	10.1

RR × DR.

Family No.	Mother's No.	Bred in pen No.	Toes.	Father's No.	Bred in pen No.	Toes.	2t.	1t.	0t.	P. ct.
42	6872	767	Normal	4562	767	Normal	0	0	6	0.0
41a	4263	767	Do	4562	767	Do	0	0	10	0.0
Totals							0	0	16	0.0

Summarizing the foregoing, and comparing the totals with Mendelian expectation, we get the result shown in table 25.

A comparison of realization and expectation in table 25 shows that the proportion of syndactyls is always less than expectation, not only for dominants and heterozygotes together, but even for pure dominants alone. The proportion of syndactyls obtained diminishes, to be sure, in accordance with expectation (on the assumption that they are pure dominants), but the numbers lag behind, in the higher proportions 40 to 25 per cent. So we reach the conclusion that, as in polydactylism, so in syndactylism dominance is very imperfect. But there is this difference, that in syndactylism dominance is so imperfect that the dominant condition rarely shows itself in heterozygotes and even fails in many pure dominants. The striking fact, the one that assures us the segregation is nevertheless occurring in this case too, is that some families (whose two parents are extracted recessives) throw 100 per cent recessives.

TABLE 25.

Nature of mating.	f	Expectation.		Realization.
		Dominants + heterozygotes.	Pure dominants.	Syndactyls.
		P. ct.	P. ct.	P. ct.
DD×DD...	41	100.0	100.0	56.1
DD×DR...	132	100.0	50.0	26.5
DR×DR...	277	75.0	25.0	10.1
RR×DR...	16	50.0	0.0	0.0
RR×RR...	119	0.0	0.0	0.0

These studies on syndactylism in poultry may be used for a critical examination of the recent work of Lewis and Embleton (1908) on syndactylism in man. The cases described by them follow the types I have just described in poultry. Their fig. 18 corresponds to my types a and α; figs. 10 and 11 to my type β. The "crossbones" referred to by the authors correspond to bones of the "curved toe." The facts presented by the authors support the idea that syndactylism is dominant rather than recessive, but they deny the application of Mendelian principles to this case. Actually, the foot deformities described by Lewis and Embleton are inherited much like syndactylism in poultry. No extracted normal (recessive) extremity produces the abnormal condition. Heterozygotes show much variation, from very abnormal to slightly abnormal (possibly perfectly normal?) appendages. Dominance is, indeed, much more potent than in poultry.

The authors' denial of the application of Mendelism to this case seems to be based on an all too superficial consideration of the hereditary behavior of the character and a tendency to "mass" statistics—a procedure that tends to obscure the interpretation of the data of heredity.

As to the inheritance of type, my statistics are not extensive enough to give a final answer, but if all types be grouped into those with straight and those with curved toes, then in crosses of straight-toed syndactyl and normal 33 per cent of the offspring were of the curved type, whereas in crosses of curved-toed syndactyls and normal 45 per cent were of the curved type. These averages depend on 22 and 15 individuals, respectively. They lead us to look for an inheritance of type when more extensive data shall be available.

Syndactylism is a typical sport, that is, a rather large mutation having a teratological aspect. The question arises, Does it prove to be prejudicial to the welfare of the species? The breeder who has only a few individuals of a rare sport feels their loss more than that of normals and the general impression left in his mind is that the sport is less capable of maintaining itself than the normal form. Assembling the data, consisting of about 40 individuals of each kind, it appears that the death-rate is not very different in the two lots; the slight excess of that of the syndactyls is sufficiently accounted for by the circumstance that no *normals* were reared during the period of greatest mortality (the summer), but were destroyed or given away as soon as hatched. It is probable, therefore, that syndactylism, under the conditions of the poultry-yard, has little life and death significance, but is one of those neutral characters whose existence Darwin clearly recognized.

CHAPTER IV.

RUMPLESSNESS.

The tail of vertebrates is, historically, the post-anal part of the trunk. Containing no longer any part of the alimentary canal, it has lost much of its primitive importance, so that its disappearance in any case is a matter of relatively little importance. Accordingly we find groups of animals in which it is rudimentary or wholly absent, such as many amphibia and the anthropoid apes and man. In all recent birds the tail is a distinct but much reduced organ—the uropygium—which contains several vertebræ in a degenerate condition. The uropygium supports the tail feathers, which are of much use in directing the bird in flight, but in ground birds, such as the grouse and poultry, seem to function only for display in the male and, in the female, to facilitate copulation.

Now, among various typically tailed vertebrates the tail is sometimes absent. Tailless dogs, cats, sheep, and horses are known; on the other hand, several cases of tails in man have been described (Harrison, 1901). Thus the tail is a part of the body subject to sporting; and it has also become the differential character for some specific groups. In other words, it is an organ that has played an important part in evolution and consequently its method of inheritance is a matter of great interest.

The origin of the tailless poultry which I have bred has been twofold. The most important strain is that referred to in an earlier report[*] as Bantam Games. The second lot consists of rumpless fowl that have arisen in my yards, spontaneously, from normal blood. Of these more later.

The two rumpless Game cocks bore the numbers 117 and 116. Dr. A. G. Phelps, of Glens Falls, New York, from whom the birds were purchased, wrote that he had imported No. 117 from England, and No. 116 was its son. The birds were very closely similar in all external features.

The matings made with No. 117 and their results are given in table 26.

TABLE 26.—*Progeny of tailless cock and tailed hens.*

Serial No.	Pen No.	Father's No.	Mother.		Offspring.			Per cent rumpless.
			No.	Races.	Condition of uropygium.			
					Present.	Small.	Absent.	
1	525	117	114	Nankin............	3	...	0	0
2	526	117	20A	Frizzle............	8	...	0	0
3	532	117	....	Bl. Coch...........	14	...	0	0
4	532a	117	127	Wh. Legh..........	19	...	0	0
4a	653	117	508	Bl. Coch.× Wh. Legh	8	3	0	0
				Totals.........	52	3	0	0

[*] Davenport, 1906, pages 62 to 64, fig. 46.

In 25 cases of the 52 an oil-gland was looked for and, in every case, it was found to be missing.

Table 26, the conclusions from which were drawn in my 1906 report, seemed to indicate the dominance of tail over its absence. On this hypothesis I suspected that if No. 117 were bred to his (tailed) offspring about 50 per cent of the progeny would be tailless, and if the tailed hybrids of the F_1 were bred together about 25 per cent of their progeny should be tailless. The actual result of such matings is shown in table 27.

TABLE 27.—*Heterozygotes mated with father.*

Serial No.	Pen No.	Tailless cock × heterozygotes.				Offspring.		
		Father.		Mother.		Condition of uropygium.		
		No.	From pen No.	No.	From pen No.	Present.	Small.	Absent.
5	653	117	Original.	577	532	6	1	0
6	653	117	..Do....	587	532	8	2	0
7	653	117	..Do....	635	532	7	0	0
8	653	117	..Do....	691	532	5	2	0
9	653	117	..Do....	652	532	15	0	0
10	653	117	..Do....	691	532	5	2	0
11	653	117	..Do....	705	532	9	2	0
12	653	117	..Do....	713	532	7	2	0
13	653	117	..Do....	760	532	13	2	0
14	653	117	..Do....	799	532	7	0	0
Total						82	13	0

TABLE 28.—*Heterozygotes mated inter se.*

Serial No.	Pen No.	Father.		Mother.		Condition of uropygium in offspring.					
						Frequency.			Percentage.		
		No.	From pen No.	No.	From pen No.	Present.	Small.	Absent.	Present	Small.	Absent.
15	661	466	526	401A	526	5	0	0	100	0	0
16	661	466	526	635	532	5	0	0	100	0	0
17	661	466	526	691	532	4	0	0	100	0	0
18	661	466	526	799	532	4	1	0	80	20	0
19	649	516	532A	521	532A	17	4	0	81	19	0
20	649	516	532A	565	532A	24	7	0	77	23	0
21	649	516	532A	665	532A	11	4	0	73	27	0
22	649	516	532A	692	532A	18	1	0	95	5	0
23	652	343	525	344	525	8	2	0	80	20	0
24	661	428	526	635	532	4	0	0	100	0	0
25	661	428	526	691	532	3	0	0	100	0	0
26	661	428	526	799	532	5	0	0	100	0	0
Total						108	19	0	85	15	0

The results given in tables 27 and 28 are remarkable. Neither in the DR × R nor the DR × DR crosses did the tail fail to develop. The tailless condition, that I had strongly suspected of being recessive and expected in 25 per cent to 50 per cent of the offspring, never once appeared. The only point of variation in the uropygium of the chicks derived from the back cross or from F_1's bred *inter se* was that in some the uropygium seemed distinctly smaller than in the others. This small uropygium was as a matter of fact recorded chiefly in chicks that failed to hatch, but it was occasionally noticed in older birds, being then usually associated with a slight convexity of the back. In some of the families the uropygium is recorded as small in suspiciously close to 25 per cent of the offspring. There is little doubt in my mind that this small uropygium represents in some way the "absence" of tail that was expected.

The next step was to cross the other rumpless bantam (No. 116), to see if he behaved like his father. Accordingly, in pen 653, I replaced the cock No. 117 by 116, the hens remaining the same, and got the result shown in table 29.

TABLE 29.—*Heterozygotes mated with No. 116.*

Serial No.	Father's No.	Mother's No.	Condition of uropygium in offspring.			
			Present.	Small.	Absent.	Per cent absent.
27	116	508	5	2	10	59
28	116	577	3	0	3	50
29	116	587	3	1	4	50
30	116	652	4	0	2	33
31	116	705	3	1	5	56
32	116	713	1	0	2	67
33	116	760	4	0	2	33
Totals (55)			23	4	28	51

Here we get a result almost exactly in accord with Mendelian expectation. Having, now, obtained rumpless hens, it became possible for the first time to test the inheritance of rumplessness in both parents. The result is shown in the table 30.

TABLE 30.—*Rumpless fowl mated inter se.*

Serial No.	Pen No.	Father.		Mother.		Condition of tail in offspring.		
		No.	From Serial No.	No.	From Serial No.	Present.	Small.	Absent.
34	742	2978	27	2601	29	0	0	4
35	854	2978	27	3430	27	0	0	9
36	742	2978	27	3430	..	*2	0	7
37	854	2978	27	2977	27	†1	0	1
Total						3	0	21

* Both from chicks that died in shell.   † From a hatched chicken.

Table 30 is unfortunately small; one may say, fragmentary. Rumpless hens are incapable of copulating unless the tail coverts are trimmed; moreover my birds have been so much inbred that they are very weak; finally, the chicks are so small that it is impracticable to rear them in brooders and the eggs are particularly apt to be broken by the brooding hens. However, it suffices to show that two tailless fowl are able to throw some tailed offspring.

The second lot of rumpless fowl, namely, those that arose *de novo* in my yards, must now be considered. In 1906, 2 birds hatched out from ordinary tailed strains. As one was a cock and the other a hen these were mated in 1907. The cock (No. 2464) came from No. 71 ♀ (a pure White Leghorn bred by myself from original White Leghorn stock described in my 1906 report) and No. 235 ♂ (an F_1 hybrid between one of these White Leghorns and my original Rose-comb Black Minorca). The hen was No. 1636. Her mother (No. 618) was an F_1 hybrid between a Minorca and Dark Brahma of series V, 1906 report, and her father (No. 637) had the same origin. Thus the parents and grandparents of both of these new rumpless birds were well known to me and known to be fully tailed and to throw only tailed birds, with the exception of these two birds.

The result of the mating of Nos. 2464 and 1636 in pen 736 was 25 chicks, of which 24 had tails and 1 (No. 5335) was without tail or oil-gland. This, unfortunately, died early, so it was impossible to breed it. In 1908, the hen No. 1636 having in the meantime died, I mated No. 2464 ♂ to 6 of his (tailed) daughters. He was not well and soon died, leaving no descendants by them, but 5 offspring by a female cousin, all tailed. Then one of his sons (tailed) was mated to its own sisters and produced 49 offspring, all tailed. Thus the strain seems to have died out. The whole history is important both because an apparently new mutation had taken place and because it was, in a degree, "hereditary."

How, if at all, can this case and those of the bantams be brought under known laws of inheritance? First of all, it must be confessed that the provisional hypothesis, suggested in my earlier report, that rumplessness is in my strain recessive has not been supported by the newer facts. In the light of the principle of imperfect dominance to which the facts of the last two chapters have led us, everything receives a satisfactory explanation. The only conclusion that meets all the facts is this: *The inhibitor of tail development—the tailless factor—is dominant; its absence—permitting a continuation of the normal development of the tail region—is recessive.*

The application of this hypothesis to the various matings may now be attempted. No. 117 is to be regarded as a heterozygote. The matings with tailed birds is of the order DR × R, and expectation in the typical case is 50 per cent DR (interrupted tail) and 50 per cent RR (non-interrupted). But, owing to the relatively weak potency of the interrupter derived from No. 117, growth of the tail is not interrupted in the heterozygous offspring. These offspring are, by hypothesis, so far as their gametes go, of two equally numerous sorts, DR and RR. Mated to No. 117, two sorts of families are to be expected, namely, the products of DR × RR (=50 per cent DR, 50 per cent RR) and the products of DR × DR (=25 per cent DD, 50 per cent DR, 25 per cent RR). The first lot of families might be expected to resemble the preceding generation in consisting entirely of tailed birds; the latter might be expected to show in the 25 per cent extracted DD's evidence of the presence of the undiluted interrupter. Actually in matings of the latter sort (table 27) 3 families show no trace of the tail-interrupter, but in 7 there is evidence of a disturbance, as shown by the small size of the uropygium and the bent back. In these families there are 13 cases of small uropygium to 53 of large, being about 20 per cent of the affected uropygium where 25 per cent was to be looked for—not a wide departure, considering the liability of not recognizing the reduced uropygium as such. This failure even of the extracted dominants completely to stop the development of the tail gives a measure of the weakness of the inhibitor in this case. Also, in table 28, matings are varied. Some are probably matings of two heterozygotes, others of two recessives, and others still of a recessive with a heterozygote. On our hypothesis we should expect some of the families of the mated hybrids to show evidence of the inhibiting factor and others to show no such evidence. In those families in which small tail appears it is found in about 19 per cent of the cases. On account of this weakness of the inhibitor in the germ-plasm of No. 117 that inhibitor is rarely fully activated. Only in one case out of the 250 or more in which that germ-plasm is used is the development of the tail completely stopped. In this case a hybrid cock derived from pen 526 (series 2, table 26) was crossed with various birds of tailed races (probable RR's), and produced in addition to 20 tailed offspring 1 devoid of uropygium and oil-gland. In this case we may conceive that an unusually potent condition of the inhibitor wholly stopped the development of the tail.

The behavior of No. 116 is that of a pure dominant. Mated to DR (and some RR?) females he produces pure dominants and heterozygotes. His inhibiting factor is potent enough to be active in the DD offspring at least; as a matter of fact 47 per cent of his get have their tails inhibited. Even in the DR's the inhibitor may sometimes work itself out. Thus No. 116 crossed on No. 508, without tailless ancestry, had 56 per cent of the progeny without tail. Since tailless birds may be either pure dominants or DR's, we may expect families of two sorts when two such are bred together— those containing only tailless offspring and those containing only 75 per cent or less of such. Both sorts of families are to be expected in a table with the composition of table 30, and both appear there.

The case of the rumpless fowl that arose *de novo* will be explained, then, as follows: Even in normal RR matings the inhibiting factor may arise by mutation. But even when two of these inhibiting factors are paired they show themselves so weak as not to appear in 25 per cent, much less the typical 75 per cent of cases, but, as in our case, merely 4 per cent. The strain takes on, indeed, the essential features of the "eversporting varieties" of De Vries (1905). It seems probable, therefore, that even in eversporting varieties inheritance may be Mendelian, modified by variations in "potency" as shown by irregularities in dominance.

CHAPTER V.

WINGLESSNESS.

The entire absence of appendages is a rare monstrosity, few cases having been cited even for man. In my experience with poultry, out of about 14,000 birds I have obtained one that had no wing on one side of the body, but this unfortunately died before being bred from. A second bird was given to me by a fancier. The bird was an Indian Game, a vigorous cock, which was handicapped by his abnormality in two ways. First, whenever he fell upon his side or back he was unable to get upon his feet without aid. On several occasions he evidently had spent hours upon the ground before he was discovered and picked up. The wings are thus clearly most important to the fowl in enabling it to regain its feet after having become prone. Secondly, he was unable to tread a hen, since this act requires the use of wings as balancers. He was, however, able to copulate with small birds without leaving the ground. Thus in two respects his abnormality would have proved fatal in nature. First, because of the personal risk, the greater since a prone bird must fall an easy prey to predaceous enemies; and secondly, because of the risk to his germ-plasm. Little wonder, then, that this abnormality should not be known among wild ground-birds.

Mated to 6 hens this wingless cock produced 130 chicks in 1907, of which all had two wings. The following year he was mated to his daughters, but died without leaving offspring. So I used a son of his to mate with his own sisters and half-sisters. The progeny in this F_2 generation consisted of 223 chicks, all of which had two wings. Thus, no trace of winglessness appeared in any of the descendants of the wingless cock.

The explanation of this case is not very certain, in view of the limited data. It seems to resemble the behavior of No. 117, the rumpless cock. And following the interpretation given in his case I would conclude that winglessness is dominant to the normal condition, that the original wingless cock was a heterozygote, and that the dominance of winglessness was imperfect in the first generation. On this hypothesis his son may well have been a pure recessive, and then all of his descendants, in turn, would be either recessives or heterozygotes (with imperfect dominance). It is, on the other hand, possible that the wingless cock was a pure dominant, but that the potency of the inhibitor was so slight as not to appear in the heterozygotes or even in extracted dominants.

CHAPTER VI.

BOOTING.

The method of inheritance of the feathering on the feet of some poultry has already been made the subject of much study. Hurst (1905, p. 152) crossed booted and non-booted birds and bred the hybrids together. He concluded that "the Mendelian principles are at work in these aberrant phenomena, but are masked by something not yet perceived." My own conclusion (1906, p. 72) was: "Booting is dominant, but usually imperfectly so." A more extended study has been desirable.

Booting is variable in amount. To indicate its degree I have had recourse to an artificial scale. I recognize 11 grades, running from 0 to 10. The grade 0 implies no feathers whatsoever. Grade 10 implies heavy booting extending over the front half of the shank. Grade 5 implies an extent of only half of the maximum, $i.\,e.$, the outer front quarter of the shank. Intermediate grades indicate intermediate extension of the feathered area.

A. TYPES OF BOOTING.

The races of booted poultry used have been as follows: First, bantam Cochins of two varieties; second, a bantam Dark Brahma; and third, the Silkie. In my representatives of the first two groups, but particularly in the Dark Brahma, the amount of booting is variable. In one type the outer third of the shank in the newly hatched chick is covered by strong, heavy, specialized feathers, directed outward, while the middle and inner thirds are covered by smaller, finer, imbricating feathers sparsely placed and resembling reduced contour-feathers. In most individuals the transition from the one kind to the other is gradual, while in others it is sharp, and in a few the outer third only of the shank is feathered. In the Silkies, which the standard poultry books describe as being more sparsely feathered on the shank,* the outer zone of feathers is the only one developed; and, occasionally, as table 31 shows, even these feathers may be lacking. We have thus two types to distinguish—the extended (Cochin, Brahma) type and the restricted type.

B. NORMAL VARIABILITY.

To appreciate the results of hybridizing we must first examine the variability of pure-blooded races. This is done in table 31.

---

* Thus Wright (1902) says the shanks of the Silkies (in England) are "slightly feathered," and Baldamus (1896) says that (in Germany) they are feathered on the *outer half*.

TABLE 31.—*Distribution of boot-grades in the offspring of Cochin, Dark Brahma, and Silkie parents.*

A. OFFSPRING OF COCHIN PARENTS.

Pen No.	Mother No.	Boot-grade.	Father No.	Boot-grade.	0	1	2	3	4	5	6	7	8	9	10	Average.
848	2297	10	545	10	..	..	..	1	..	..	1	..	..	1	18	9.43
776	2574	10	2732	8	..	..	..	..	..	..	..	..	3	2	6	9.27
848	2300	8	545	10	..	..	..	..	..	..	..	..	1	2	5	9.25
776	2570	6	2732	8	..	..	..	..	..	..	1	1	..	11	1	8.71
848	2075	9	545	10	..	..	..	..	..	1	1	..	..	..	4	8.50
776	2072	6	2732	8	..	..	..	..	..	..	1	..	4	2	2	8.44
758	130	6	545	10	..	..	*..	..	..	..	1	1	3	9	..	8.20
776	2073	6	2732	8	..	..	..	..	..	1	2	..	2	10	1	8.00
776	2300	6	2732	8	..	..	..	..	..	1	..	1	3	6	5	2	8.00
758	131	10	545	10	..	..	..	..	..	1	..	4	6	1	1	7.96
776	2297	6	2732	8	..	..	..	1	..	1	..	3	6	6	2	7.95
776	1132	3	2732	8	..	..	..	1	1	1	1	3	6	8	..	7.57
776	2937	7	2732	8	..	..	..	..	..	..	..	1	3	3	1	7.50
776	2299	7	2732	8	..	..	1	..	..	1	1	4	7	3	1	7.44
Totals (199)					..	..	1	3	3	8	9	24	47	61	43	8.24

B. OFFSPRING OF DARK BRAHMA PARENTS.

[All individuals have sprung from No. 121 ♀ (boot of grade 9) and No. 122 ♂ (boot of grade 6).]

Pen No.	Mother No.	Boot-grade.	Father No.	Boot-grade.	0	1	2	3	4	5	6	7	8	9	10	Average.
816	2030	6	122	6	..	..	..	..	..	..	..	..	..	1	3	9.8
703	2030	6	122	6	..	..	..	..	..	..	4	2	0	3	6	8.3
816	121	9	122	6	..	..	..	..	..	1	3	1	2	4	5	8.3
816	5979	6	122	6	..	..	..	..	..	..	1	0	2	..	..	7.3
816	2353	5	122	6	..	..	..	*1	..	*1	1	1	0	1	0	2	7.1
816	5835	5	122	6	..	..	..	*1	0	1	2	..	..	1	3	6.5
816	5840	5	122	6	..	..	..	*1	..	..	1	..	..	..	1	6.3
703	2353	5	122	6	..	..	..	..	1	1	3	..	1	..	..	5.8
Totals (61)					..	..	..	2	2	4	15	3	6	9	20	7.62

C. OFFSPRING OF SILKIE PARENTS.

Pen No.	Mother No.	Boot-grade.	Father No.	Boot-grade.	0	1	2	3	4	5	6	7	8	9	10	Average.
734	468	4	774	3	..	..	1	2	..	..	1	1	..	..	..	4.20
734	1002	3	774	3	..	..	1	4	..	1	3	..	..	..	..	4.11
734	841	(?)	774	3	..	..	..	..	2	..	..	..	..	..	..	4.00
815	7434	7	774	3	..	..	..	..	2	..	..	..	..	..	..	4.00
734	773	1	774	3	..	..	..	2	2	..	..	..	..	..	..	3.50
734	680	1	774	3	..	..	..	2	..	..	..	..	..	..	..	3.00
734	405A	1	774	3	..	..	1	3	1	..	..	..	..	..	..	3.00
815	499	2	774	3	1	1	3	..	..	2	..	1	..	..	..	3.00
734	499	2	774	3	1	1	5	2	2	1	..	..	..	..	..	2.50
734	500	1	774	3	2	1	2	3	..	..	..	..	..	..	..	1.75
815	773	1	774	3	4	1	3	..	..	..	..	..	..	..	..	1.25
815	500	1	774	3	1	1	..	..	..	..	..	..	..	..	..	0.50
815	496	3	774	3	1	..	..	..	..	..	..	..	..	..	..	0.00
Totals (68)					10	5	16	18	9	4	4	2	..	..	..	2.72

SUMMARY.

Races.	0	1	2	3	4	5	6	7	8	9	10	Average.
Cochins	..	..	0.5	1.5	1.5	4.0	4.5	12.1	23.6	30.7	21.6	8.24
Dark Brahmas	..	..	..	3.3	3.3	6.6	24.6	4.9	9.8	14.8	32.8	7.62
Silkie	14.8	7.4	23.5	26.5	13.2	5.9	5.9	2.9	..	..	..	2.72

* Determination made on embryo chicks.

An inspection of table 31 shows that, in respect to booting, the Cochins and Dark Brahmas are clearly closely related to each other. Owing to smaller numbers and to other circumstances that will be discussed later, the results are less regular in the Dark Brahma offspring, but in both the range is from 2 or 3 upward to 10, with a great preponderance in grades above 5. In the Silkies, on the other hand, the greatest frequency is found in grades below 5. This difference is correlated with a difference of the parents, for the commonest grades of the parents of the Cochins are between 6 and 10, of the Dark Brahmas between 5 and 9, and of the Silkies between 1 and 3. These results suggest that the Silkie is typically heterozygous in boot, producing 25 per cent recessives (boot of grade 4–7) and 75 per cent dominant (0, 1) and heterozygous (2, 3). We shall see that this hypothesis receives support from all Silkie matings.

Inside of any part of this table it appears that, on the whole, as the average grade of the boot in the progeny diminishes that of the parentage diminishes, although the correlation is by no means perfect. Thus the average of the parental grades in the first part of table 31, A (which is arranged in descending order of the averages of the offspring) is 8.5; in the lower half, 7.4. The average of parental grades in the upper half of table 31, B is 6.4; in the lower half 5.5. In table 31, C the grades are 2.9 and 2.3, respectively. This correlation indicates, without exactly measuring, heredity in grade of booting.

Table 32 shows the results of crosses between Cochins (high grade of boot) and Silkies (low grade).

TABLE 32.—*Distribution of boot-grades between a high and low grade of boot in parents.*

Pen No.	Mother.				Father.				Grade of boot in offspring.											
	No.	Gen.	Races.	Grs.	No.	Gen.	Race.	Grs.	0	1	2	3	4	5	6	7	8	9	10	Average.
851	5567	P	Bl.×Bf. C...	9	7526	P	Silkie......	3	..	..	..	..	2	..	..	..	3	3	5	8.15
851	3410	P	...Do......	9	7526	P	...Do......	3	..	..	..	..	..	4	3	2	1	6	1	7.29
851	6956	P	...Do......	8	7526	P	...Do......	3	..	..	..	..	3	3	..	2	2	..	5	7.13
851	2073	P	...Do......	7	7526	P	...Do......	3	..	1	..	1	1	..	1	1	1	3	2	6.91
851	2299	P	...Do......	7	7526	P	...Do......	3	..	..	..	..	2	2	1	1	..	..	3	6.78
851	840	P	Bf. C......	10	7526	P	...Do......	3	..	..	..	..	1	..	1	..	..	1	..	6.33
851	1002	P	...Do......	8	7526	P	...Do......	3	..	..	..	3	1	2	1	2	4	1	1	6.27
815	131	P	Bk. C......	10	774	P	...Do......	4	..	..	..	3	1	1	2	2	1	1	2	6.23
851	841	P	Bf. C	10	7526	P	...Do......	3	..	..	..	..	..	1	..	1	..	1	..	6.00
851	838	P	...Do......	8	7526	P	...Do......	3	..	..	..	4	2	4	3	..	..	2	2	5.65
Totals (116)................									0	1	0	11	14	16	13	10	13	17	21	6.77

So far as the average grade of boot in offspring goes, this table stands between that of the Cochins (table 31, A) and that of the Silkies (table 31, C). But what is especially striking is the apparent dimorphism revealed in the line of totals. There is one (empirical) mode at 10, corresponding with that of the Cochins, and a second clear mode at 5, corresponding to that of the Silkies. If we assume the Cochin to be homozygous in boot (RR) and the Silkie to be heterozygous in boot, then we can interpret the high mode as extracted recessives, the median mode as heterozygotes.

C. RESULTS OF HYBRIDIZATION.

We have next to consider the nature of the inheritance when one parent belongs to an unbooted race, the other to a booted one (table 33).

TABLE 33.—*Distribution of boot-grades in the F_1 generation of booted × non-booted parents.*

A. COCHIN CROSSES.

Pen No.	Mother.				Father.				Grade of boot in offspring.											
	No.	Gen.	Races.	Grs.	No.	Gen.	Races.	Grs.	0	1	2	3	4	5	6	7	8	9	10	Average.
773	1334	P	W. Legh.	0	836	P	Bl. Coch	10	..	..	..	3	1	1	1	1	..	2	..	5.44
773	193	P	...Do.	0	836	P	...Do.	10	..	1	2	6	8	7	4	2	..	..	..	4.37
773	1366	P	...Do.	0	830	P	...Do.	10	..	..	..	2	5	2	1	..	..	..	..	4.20
773	127	P	...Do.	0	836	P	...Do.	10	..	..	3	10	9	12	4	..	..	..	..	4.11
773	692	P	W.Legh.(R)	0	836	P	...Do.	10	..	..	..	10	3	2	..	..	..	..	..	3.47
774	2075	P	Coch.	8	1431	P	W.Legh.(R)	0	6	1	1	..	1	..	..	..	..	..	..	0.78
Totals (111)									6	2	6	31	27	24	10	3	0	2	0	3.91

B. DARK BRAHMA CROSSES.

Pen No.	No.	Gen.	Races.	Grs.	No.	Gen.	Races.	Grs.	0	1	2	3	4	5	6	7	8	9	10	Average.
727	Y	P	D. Br.	10	381	P	Houd.	0	..	..	..	..	2	3	2	1	2	..	..	5.80
727	121	P	...Do.	10	381	P	...Do.	0	1	..	..	1	1	5	4	..	..	..	..	4.67
823	2030	P	...Do.	7	3858	P	M×P.	0	..	..	5	16	15	4	1	2	..	..	..	3.67
823	Y	P	...Do.	8	3858	P	...Do.	0	..	..	1	7	6	2	..	..	..	..	..	3.56
838	3814	P	W. Legh.	0	122	P	D. Br.	6	..	2	2	6	6	1	1	..	..	..	..	3.28
838	202	P	Min.	0	122	P	...Do.	6	..	..	2	5	3	..	..	..	..	..	..	3.10
838	71	P	W. Legh.	0	122	P	...Do.	6	..	..	..	1	..	1	..	..	..	..	..	3.00
838	3832	P	...Do.	0	122	P	...Do.	6	1	1	..	1	1	2	..	..	..	..	..	3.00
838	10	P	...Do.	0	122	P	...Do.	6	..	1	..	3	1	..	..	..	..	..	..	2.80
816	121	P	D. Br.	9	4912	P	M×P.	0	..	..	8	4	1	1	..	..	..	..	..	2.64
816	5838	P	...Do.	9	4912	P	...Do.	0	..	..	5	5	1	..	..	..	..	..	..	2.64
838	5418	P	W.L., Min.	0	122	P	D. Br.	6	1	1	3	3	1	1	..	..	..	..	..	2.50
816	5979	P	D. Br.	6	4912	P	M×P.	0	4	3	4	7	4	1	1	..	..	..	..	2.46
816	2353	P	...Do.	5	4912	P	...Do.	0	..	2	2	4	1	..	..	..	..	..	..	2.44
816	5977	P	...Do.	4	4912	P	...Do.	0	..	3	2	1	..	1	..	..	..	..	..	2.14
816	5835	P	...Do.	5	4912	P	...Do.	0	3	5	5	8	3	..	..	..	..	..	..	2.13
816	5840	P	...Do.	5	4912	P	...Do.	0	5	1	3	4	1	..	..	..	..	..	..	1.64
823	6626	P	...Do.	2	3858	P	...Do.	0	1	10	2	2	..	..	..	..	..	..	..	1.33
816	5980	P	...Do.	5	4912	P	...Do.	0	5	8	1	5	..	..	..	..	..	..	..	1.33
Totals (268)									21	37	45	83	47	21	9	3	2	0	0	2.84

C. SILKIE CROSSES.

Pen No.	No.	Gen.	Races.	Grs.	No.	Gen.	Races.	Grs.	0	1	2	3	4	5	6	7	8	9	10	Average.
744	777	P	Silkie.	8	1176	P	W. Legh.	0	3	..	1	1	1	..	..	..	..	..	..	1.80
744	681	P	...Do.	5	1176	P	...Do.	0	11	2	1	1	1	1	..	..	..	..	..	0.94
744	469	P	...Do.	1	1176	P	...Do.	0	11	3	..	..	..	..	..	..	..	..	..	0.21
Totals (37)									25	5	2	2	2	1	0	0	0	0	0	0.76

SUMMARY.

Crosses.	Grades of boot in offspring, reduced to percentages.											
	0	1	2	3	4	5	6	7	8	9	10	Average.
Cochin	5.4	1.8	5.4	28.0	24.3	21.6	9.0	2.7	0.0	1.8	...	3.91
Brahma	7.8	13.8	16.8	31.0	17.5	7.8	3.4	1.1	0.7	...	...	2.84
Silkie	67.6	13.5	5.4	5.4	5.4	2.7	...	...	...	...	...	0.76

An inspection of Table 33, which gives the distribution of grades of boot in the offspring constituting the first hybrid generation, might well lead to the conclusion that inheritance is here of a blending nature, or that, if either condition is dominant, it is the booted one, as suggested in my report of 1906. On this hypothesis the offspring with no boot illustrate imperfection of dominance, and one would say that, in booting, dominance is very imperfect.

However plausible such an interpretation might appear when based on the first hybrid generation alone, it becomes untenable when subsequent generations are taken into account, as we shall see later. The hypothesis breaks down completely in the second hybrid generation and we are forced to the opposite hypothesis, namely, that the clean-shanked condition is dominant. Such an hypothesis would seem, at first, to contravene the principle enunciated in my report of 1906 that the more progressive condition is dominant over the less progressive condition, or absence. But such is not necessarily the fact. We have no right to assume that presence of boot is the new character. The rest of the body of poultry (save the head) is covered with feathers. If the foot is not it must be because there is something in the skin of the foot that inhibits the development of feathers there. And this inhibiting factor is dominant over its absence.

Table 33 shows that the Silkie crosses yield an exceptionally high per cent of the dominant clear-footed condition. This is additional evidence that the Silkies are DR, and so this cross produces 50 per cent of pure extracted dominants in addition to 50 per cent of heterozygotes in booting.

To get further light on the nature of inheritance of booting we pass to the examination of the second hybrid generation (table 34).

In the case of Silkies, which throw 67.6 per cent clean-shanked progeny in F_1, we find in F_2 only about 60 per cent clean-shanked. This diminution is, of course, due to the extraction of some pure booted recessives, which draw from the proportion of clean shanks.

In the case of the Cochins and Dark Brahmas, expectation, with perfect dominance, is that 75 per cent of the offspring shall be clean-shanked. Since dominance is imperfect (as shown by the occurrence of many booted birds in F_1) we should look for an actual failure to reach so large a proportion, but we are hardly prepared for the result that in most of the F_2 crosses of Cochins and Brahmas less than 25 per cent of the offspring are clean-shanked. In 4 pens the average is only 10 to 12 per cent, and in one only 2 per cent of the offspring fail to develop feathers on the feet. What shall we say of such a case as the last? The history of the father (No. 666) is absolutely certain; his mother was No. 121, the original Dark Brahma female, with a boot of grade 9 and a record in her immediate progeny that indicates perfect purity of booting in her germ-cells. His father was a White Leghorn with clean shanks and without a suspicion of having such antipodal blood as the Asiatic in his ancestry. No. 666 is certainly heterozygous in boot, if boot is a single unit. The hens with which No. 666 were mated were clearly heterozygous, as is known not only from their ancestry, but also from their behavior when mated with another cock, No. 254, in which case they threw 12 per cent non-booted offspring. If now both parents are heterozygous they must produce 25 per cent recessives. This is the fact that forces us to conclude that clean shank is not recessive, but dominant and due to an inhibitor that frequently *fails to dominate*. In table 31 the two recessive varieties, mated *inter se*, produce no featherless shanks; the feathers grow freely as they do over the rest of the body. Some of the Silkies of table 31, however, are really heterozygous, with the dominant inhibitor not showing; consequently they throw a large proportion of non-booted offspring. In F_1, as table 33 shows, the heterozygous offspring have a reduced boot and perfect dominance—complete inhibition of boot—in from 6 to 68 per cent. Dominance is most complete in the Silkies, where, the feathering being feeble, the inhibitor has, as it were, less to do in overcoming it. In F_2 the expected 75 per cent dominant is approached in the case of the Silkies (62 per cent and 59 per cent, respectively), but inhibition is very imperfect in the Cochin and Brahma crosses, being reduced to between 25 and 2 per cent. More proof that boot is due to the absence of a factor rather than to its presence is found in this generation. If absence of boot is recessive, then, combined with imperfection of dominance, *at least* 25 per cent of the offspring should be recessive and probably a much larger proportion. The results in table 34 are absolutely incompatible with this hypothesis, since, in one case, there are only 2 per cent that can not develop boot. Two extracted clean-footed birds sometimes throw boot and sometimes not, and this result is to be expected on the hypothesis that clean-footedness is dominant, but two heavily booted birds can not transmit the boot inhibitor.

TABLE 34.—*Distribution of boot-grade in the F_2 generation of booted × non-booted poultry.*

Pen No.	Mother.				Father.				Offspring.			
	No.	Gen.	Races.	Grade.	No.	Gen.	Races.	Grade.	Boot present.	Boot slight.	Boot absent.	P. ct. absent.
COCHIN CROSSES.												
650	170	F_1	Bl. Coch.×Wh. Legh......	Pr.	265	F_1	Bl. Coch.×Wh. Legh.......	Pr.	19	2	2	8.7
650	263	F_1	....Do.....................	Pr.	265	F_1	....Do.....................	Pr.	36	2	2	5.0
650	278	F_1	....Do.....................	Pr.	265	F_1	....Do.....................	Pr.	26	4	4	11.8
650	361	F_1	....Do.....................	Pr.	265	F_1	....Do.....................	Pr.	24	2	9	25.7
650	364	F_1	....Do.....................	Pr.	265	F_1	....Do.....................	Pr.	39	5	3	6.4
			Totals (179)..............						144	15	20	11.1
654	602	F_1	Wh. Legh.×Bf. Coch......	Pr.	704	F_1	Wh. Legh.×Bf. Coch.......	Pr.	11	4	5	25.0
654	828	F_1	....Do.....................	Pr.	704	F_1	....Do.....................	Pr.	7	11	0	0.0
654	640	F_1	....Do.....................	Pr.	704	F_1	....Do.....................	Pr.	13	2	3	16.7
654	696	F_1	....Do.....................	Pr.	704	F_1	....Do.....................	Pr.	8	5	8	38.1
654	767	F_1	....Do.....................	Pr.	704	F_1	....Do.....................	Pr.	3	1	3	42.9
654	697	F_1	....Do.....................	Pr.	704	F_1	....Do.....................	Pr.	4	3	6	46.2
			Totals (97)...............						46	26	25	25.8

BOOTING.

TABLE 34.—*Distribution of boot-grade in the F_2 generation of booted × non-booted poultry*—Continued.

DARK BRAHMA CROSSES.

Pen No.	Mother.				Father.				Offspring.			
	No.	Gen.	Races.	Grade.	No.	Gen.	Races.	Grade.	Boot present.	Boot slight.	Boot absent.	P. ct. absent.
608	384	F_1	Wh. Legh. × Dk. Brah.	Pr.	409	F_1	Wh. Legh. × Dk. Brah.	Pr.	36	5	3	6.8
608	248	F_1	...Do.	Pr.	409	F_1	...Do.	Pr.	32	5	4	9.8
608	249	F_1	...Do.	Pr.	409	F_1	...Do.	Pr.	39	11	13	20.6
608	395	F_1	...Do.	Pr.	409	F_1	...Do.	Pr.	20	11	10	24.4
608	385	F_1	...Do.	Pr.	409	F_1	...Do.	Pr.	20	6	14	35.0
			Totals (229)						147	38	44	19.2
659	762	F_1	Wh. Legh. × Dk. Brah.	Pr.	375	F_1	Wh. Legh. × Dk. Brah.	Pr.	18	4	1	4.4
659	503	F_1	...Do.	Pr.	375	F_1	...Do.	Pr.	23	6	2	6.5
659	382	F_1	...Do.	Pr.	375	F_1	...Do.	Pr.	10	2	1	7.7
659	250	F_1	...Do.	Pr.	375	F_1	...Do.	Pr.	33	7	5	11.1
659	737	F_1	...Do.	Pr.	375	F_1	...Do.	Pr.	19	2	3	12.5
659	387	F_1	...Do.	Pr.	375	F_1	...Do.	Pr.	16	6	4	15.4
			Totals (162)						119	27	16	9.9
655	720	F_1	Wh. Legh. × Dk. Brah.	Pr.	666	F_1	Wh. Legh. × Dk. Brah.	Sl.	5	2	..	0.0
655	724	F_1	...Do.	Pr.	666	F_1	...Do.	Sl.	6	1	..	0.0
655	728	F_1	...Do.	Pr.	666	F_1	...Do.	Sl.	3	1	..	0.0
655	730	F_1	...Do.	Pr.	666	F_1	...Do.	Sl.	4	..	..	0.0
655	732	F_1	...Do.	Pr.	666	F_1	...Do.	Sl.	9	..	..	0.0
655	734	F_1	...Do.	Pr.	666	F_1	...Do.	Sl.	3	..	..	0.0
655	761	F_1	...Do.	Pr.	666	F_1	...Do.	Sl.	6	2	..	0.0
655	800	F_1	...Do.	Pr.	666	F_1	...Do.	Sl.	1	..	..	0.0
655	721	F_1	...Do.	Pr.	666	F_1	...Do.	Sl.	9	1	1	9.1
			Totals (54)						46	7	1	1.9
655	724	F_1	Wh. Legh. × Dk. Brah.	Pr.	254	F_1	Wh. Legh. × Dk. Brah.	Pr.	3	..	..	0.0
655	734	F_1	...Do.	Pr.	254	F_1	...Do.	Pr.	12	1	..	0.0
655	800	F_1	...Do.	Pr.	254	F_1	...Do.	Pr.	13	..	1	7.1
655	720	F_1	...Do.	Pr.	254	F_1	...Do.	Pr.	12	..	1	7.7
655	728	F_1	...Do.	Pr.	254	F_1	...Do.	Pr.	8	1	1	10.0
655	761	F_1	...Do.	Pr.	254	F_1	...Do.	Pr.	17	4	4	16.0
655	732	F_1	...Do.	Pr.	254	F_1	...Do.	Pr.	8	1	2	18.2
655	730	F_1	...Do.	Pr.	254	F_1	...Do.	Pr.	7	..	2	22.2
655	721	F_1	...Do.	Pr.	254	F_1	...Do.	Pr.	9	..	3	25.0
			Totals (110)						89	7	14	12.7
632	742	F_1	Min. × Dk. Brah.	Pr.	637	F_1	Min. × Dk. Brah.	Pr.	4	1	0	0.0
632	690	F_1	...Do.	Pr.	637	F_1	...Do.	Pr.	27	6	1	2.9
632	631	F_1	...Do.	Pr.	637	F_1	...Do.	Pr.	32	11	2	4.4
632	618	F_1	...Do.	Pr.	637	F_1	...Do.	Pr.	35	8	2	4.4
632	700	F_1	...Do.	Pr.	637	F_1	...Do.	Pr.	18	3	2	8.7
632	703	F_1	...Do.	Pr.	637	F_1	...Do.	Pr.	14	11	3	10.7
632	743	F_1	...Do.	Pr.	637	F_1	...Do.	Pr.	22	2	3	11.1
632	599	F_1	...Do.	Pr.	637	F_1	...Do.	Pr.	23	8	4	11.4
632	524	F_1	...Do.	Pr.	637	F_1	...Do.	Pr.	18	6	5	17.2
632	576	F_1	...Do.	Pr.	637	F_1	...Do.	Pr.	14	9	6	20.7
632	638	F_1	...Do.	Pr.	637	F_1	...Do.	Pr.	8	2	6	37.5
			Totals (316)						215	67	34	10.8

Pen No.	Mother.				Father.				Boot-grade in offspring.											Average.	P. ct. absent.
	No.	Gen.	Races.	Gr.	No.	Gen.	Races.	Gr.	0	1	2	3	4	5	6	7	8	9	10		
801	2526	F_1	Min. × Dk. Brah.	2	5399	F_1	W. L. × Dk. Brah.	8	..	..	..	..	1	..	..	1	..	..	1	7.0	0.0
801	2831	F_1	...Do.	4	5399	F_1	...Do.	8	1	1	1	4	1	7	2	2	2	..	2	5.0	4.3
801	1892	F_1	...Do.	3	5399	F_1	...Do.	8	1	1	0	1	2	..	1	..	1	1	1	5.0	11.1
			Totals (35)						2	2	1	5	4	7	3	3	3	1	4	5.2	5.71

50 INHERITANCE OF CHARACTERISTICS IN DOMESTIC FOWL.

TABLE 34.—*Distribution of boot-grade in the F_2 generation of booted × non-booted poultry*—Continued.

SILKIE CROSSES.

Pen No.	Mother.				Father.				Boot-grade in offspring.											Average.	P. ct. absent.
	No.	Gen.	Races.	Gr.	No.	Gen.	Races.	Gr.	0	1	2	3	4	5	6	7	8	9	10		
709	1955	F₁	Silkie×Spanish.	5	1578	F₁	Silkie×Spanish.	0	5	1	2	1	1	1	1	0	0	0	0	1.92	41.7
753	1966	F₁	Silkie×Min.	0	2573	F₁	Min.×Silkie.	0	19	4	2	2	..	2	2	1	..	..	1.71	55.9
709	1963	F₁	Silkie×Spanish.	7	1578	F₁	Silkie×Spanish.	0	23	6	1	6	7	..	..	..	..	..	1.26	53.5
753	2575	F₁	Silkie×Min.	0	2573	F₁	Silkie×Min.	0	15	3	7	..	..	..	..	..	..	..	0.68	60.0
753	2071	F₁	....Do........	0	2573	F₁	....Do........	0	23	4	6	..	..	..	..	..	..	..	0.49	69.7
709	1453	F₁	....Do........	1	1578	F₁	Silkie×Spanish.	0	24	11	3	..	..	..	..	..	..	..	0.45	63.2
709	2223	F₁	Silkie×Spanish.	0	1578	F₁	....Do........	0	32	7	3	..	..	..	..	..	..	..	0.31	76.2
			Totals (227)						141	36	24	9	8	3	3	2	1	0	0	0.87	62.2
830	4082	F₁	Silkie×W. Legh.	2	3947	F₁	Silkie×W. Legh.	1	11	8	..	7	1	..	..	..	..	..	1.22	40.7
830	4079	F₁	....Do........	0	3947	F₁	....Do........	1	18	7	6	3	..	..	..	..	..	..	0.82	53.0
830	5379	F₁	....Do........	0	3947	F₁	....Do........	1	18	4	5	3	..	..	..	..	..	..	0.77	60.0
830	4081	F₁	....Do........	0	3947	F₁	....Do........	1	24	6	10	1	..	..	..	..	..	..	0.71	58.5
830	5374	F₁	....Do........	0	3947	F₁	....Do........	1	11	3	3	1	..	..	..	..	..	..	0.67	61.1
830	3946	F₁	....Do........	0	3947	F₁	....Do........	1	19	1	..	..	1	..	..	..	..	..	0.24	90.5
			Totals (170)						101	29	24	14	2	0	0	0	0	0	0.75	59.4

The distribution of table 35 is characterized by its large variability. Although the numbers are small, there are evidences of two modes, one between grades 3 and 6, and the other at from 8 to 10; these evidently correspond to the modes of the typical Silkie and the typical Cochin respectively or to DR and RR types of booting respectively. The distribution of table 35 is additional evidence of the heterozygous nature of the Silkie boot.

TABLE 35.—*Distribution of boot-grades in Silkie × Cochin crosses.*

Pen No.	Mother.				Father.				Boot-grades in offspring.											Av.	P. ct. abs.
	No.	Gen.	Races.	Grs.	No.	Gen.	Races.	Grs.	0	1	2	3	4	5	6	7	8	9	10		
821	5925	F₁	Silk.×Coch.	7	6095	F₁	Silk.×Coch.	7	..	..	..	..	1	..	..	1	3	1	1	7.7	0.0
821	7408	F₁	....Do......	4	6095	F₁	....Do......	7	..	..	..	1	2	2	3	..	2	1	2	6.5	0.0
821	7413	F₁	....Do......	3	6095	F₁	....Do......	7	2	0	3	1	0	1	1	0	0	1	1	3.9	20.0
821	7416	F₁	....Do......	5	6095	F₁	....Do......	7	..	..	..	3	1	0	4	0	3	3	2	6.8	0.0
821	7417	F₁	....Do......	..	6095	F₁	....Do......	7	..	..	..	..	..	..	1	..	..	1	4	9.3	0.0
821	7418	F₁	....Do......	4	6095	F₁	....Do......	7	..	..	1	..	2	1	1	1	1	..	1	5.8	0.0
821	7423	F₁	....Do......	6	6095	F₁	....Do......	7	..	..	..	1	..	2	..	2	2	..	2	7.0	0.0
821	7428	F₁	....Do......	..	6095	F₁	....Do......	7	1	..	..	..	..	1	..	..	1	..	..	4.3	33.3
821	7429	F₁	....Do......	8	6095	F₁	....Do......	7	..	..	..	1	1	1	..	..	..	1	1	6.2	0.0
			Totals (77)						3	0	4	7	7	8	9	5	12	8	14	6.42	3.90
											29						48					

We are now in a position to consider the effect of back crosses (table 36). The contrast between the totals in tables 36 and 37 is very great. The strict Mendelian expectation is: in the DR × D crosses 50 per cent DD (clean-footed) and 50 per cent heterozygous, which, with imperfect dominance, might be expected to show foot-feathering. Actually about 46 per cent are clean-footed. In the DR × R crosses expectation is that 50 per cent certainly (the extracted recessives) and 50 per cent more possibly will have the shanks feathered, on account of imperfect dominance of the heterozygotes. Actually all have feathered feet. These statistics thus confirm the view of the dominance of the inhibiting factor. Were clean shank recessive, then the DR × R crosses must give 50 per cent clean-footed and probably over. The actual result, none clean-footed, is not in accord with the latter assumption.

TABLE 36.—*Distribution of boot-grade in DR×D (non-booted) crosses.*

Pen No.	Mother.				Father.				Boot-grade in offspring.			
	No.	Gen.	Races.	Grade.	No.	Gen.	Race.	Grade.	Present.	Slight.	Absent.	Per cent. present.
653	508	F_1	Wh. Legh.×Bf. Coch...	Pr.	117	P.	Game...	0	3	4	6	46.2
653	508	F_1	...Do..............	Pr.	116	P.	...Do...	0	6	5	4	26.7
653	577	F_1	R×Bf. Coch..........	3	117	P.	...Do...	0	1	0	7	87.5
653	577	F_1	....Do..............	3	116	P.	...Do...	0	1	3	2	33.3
653	587	F_1	....Do..............	1	117	P.	...Do...	0	1	2	4	57.1
653	587	F_1	....Do..............	1	116	P.	...Do...	0	3	3	2	25.0
653	635	F_1	....Do..............	3	117	P.	...Do...	0	..	1	6	85.7
653	635	F_1	....Do..............	3	116	P.	...Do...	0	2	2	1	20.0
653	652	F_1	....Do..............	5	117	P.	...Do...	0	5	8	4	23.5
653	652	F_1	....Do..............	5	116	P	...Do...	0	1	2	2	40.0
653	691	F_1	....Do..............	Pr.	117	P.	...Do...	0	2	2	1	20.0
653	705	F_1	....Do..............	2	117	P.	...Do...	0	3	2	5	50.0
653	705	F_1	....Do..............	2	116	P.	...Do...	0	1	1	5	71.4
653	713	F_1	....Do..............	Pr.	117	P	...Do...	0	..	0	4	100.0
653	713	F_1	....Do..............	Pr.	116	P.	...Do...	0	1	1	3	60.0
653	760	F_1	....Do..............	Pr.	117	P.	...Do...	0	2	2	6	60.0
653	760	F_1	....Do..............	Pr.	116	P.	...Do...	0	0	3	2	40.0
653	799	F_1	....Do..............	3	117	P.	...Do...	0	2	0	3	60.0
	Total (143)								34	42	67	46.9
661	635	F_1	Bf. Coch.×Game......	Pr.	466	P.	Game...	0	1	..	2	66.7
661	635	F_1	....Do..............	Pr.	428	P.	...Do...	0	2	..	1	33.3
661	691	F_1	....Do..............	Pr.	466	P.	...Do...	0	2	..	2	50.0
661	691	F_1	....Do..............	Pr.	428	P.	...Do...	0	2	..	1	33.3
661	799	F_1	....Do..............	Pr.	466	P.	...Do...	0	3	..	2	40.0
661	799	F_1	....Do..............	Pr.	428	P.	...Do...	0	4	..	1	20.0
	Total (23)								14	0	9	39.1
	Grand total (166)..............................								48	42	76	45.8

TABLE 37.—*Distribution of boot-grade in DR × RR (booted) crosses.*

Pen No.	Mother.				Father.				Boot-grade in offspring.										
	No.	Gen.	Race.	Gr.	No.	Gen.	Race.	Gr.	0	1	2	3	4	5	6	7	8	9	10
851	838	P.	Cochin........	8	7526	*F_1	Silkie......	3	..	..	..	3	2	4	3	..	..	2	2
851	840	P.	....Do........	10	7526	F_1	....Do.....	3	..	..	..	..	1	..	1	..	..	1	..
851	841	P.	....Do........	10	7526	F_1	....Do.....	3	..	..	..	..	1	..	1	..	1	..	1
851	1002	P.	....Do........	8	7526	F_1	....Do.....	3	..	..	..	3	1	2	1	2	3	1	1
851	2073	P.	....Do........	7	7526	F_1	....Do.....	3	..	..	1	1	1	1	1	1	1	3	2
851	2299	P.	....Do........	9	7526	F_1	....Do.....	3	..	..	..	..	2	2	1	1	..	..	2
851	3410	P.	....Do........	9	7526	F_1	....Do.....	3	..	..	..	..	..	4	3	2	1	5	1
851	5567	P.	....Do........	9	7526	F_1	....Do.....	3	..	..	..	..	2	..	..	..	3	3	5
851	6956	P.	....Do........	8	7526	F_1	....Do.....	3	..	..	..	..	3	3	..	2	2	..	5
	Totals (99)........................								0	0	1	7	13	15	11	8	11	15	18

* Pure-blooded Silkie assumed heterozygous in boot.

Numerous observations have been made upon the progeny of parents belonging to hybrid generations beyond the first. Owing to the extreme imperfection of dominance it is rarely possible to say with certainty from inspection whether a given bird has germ-cells dominant or recessive, or

52 INHERITANCE OF CHARACTERISTICS IN DOMESTIC FOWL.

both, with reference to booting; only breeding enables us to make a decision. There is an exception, however, in the case of pure extracted recessives. They are distinguished by heavy booting and produce only booted offspring. I propose to give, in detail, the matings of these later generations and their progeny, the families being arranged in decreasing order of average grade of booting (table 38).

TABLE 38.—*Distribution of boot-grades in offspring of parents one or both of which belong to a hybrid generation beyond the first.*

B = Brahma; C = Cochin; G = Game; L = Leghorn; M = Minorca; S = Silkie; Sp = Spanish; T = Tosa; WL = White Leghorn

Serial No.	Pen No.	Mother.				Father.				Mating.	Boot-grade in offspring.											
		No.	Gen.	Races.	Gr.	No.	Gen.	Races.	Gr.		0	1	2	3	4	5	6	7	8	9	10	Av.
1	814	354	F_1	B×T	7	3975	F_2	B×T	9	R×R	..	..	..	..	..	..	..	..	..	10	15	9.6
2	801	181	F_1	..Do	4	5399	F_2	M×B	8	.Do	..	..	..	..	..	..	..	..	..	1	1	9.5
3	814	300	F_1	..Do	5	3975	F_2	B×T	9	.Do	..	..	..	..	..	..	..	..	1	3	4	9.4
4	801	4569	F_2	..Do	6	4562	F_2	..Do	7	.Do	..	..	..	..	..	..	..	..	1	1	2	9.3
5	814	5523	F_2	..Do	9	3975	F_2	..Do	9	.Do	..	..	..	..	..	..	1	..	..	3	4	9.1
6	814	4560	F_2	..Do	8	3975	F_2	..Do	9	.Do	..	..	..	..	..	1	1	1	..	2	7	8.8
7	814	190	F_1	..Do	2	3975	F_2	..Do	9	.Do	..	..	..	..	..	..	1	1	1	1	4	8.8
8	806	4325	F_2	M×B	7	5257	F_2	M×B	9	.Do	..	..	..	..	..	..	..	1	1	2	3	8.6
9	806	5913	F_2	..Do	7	5257	F_2	..Do	9	.Do	..	..	..	..	1	..	..	1	4	2	3	8.3
10	732	1235	F_2	..Do	8	2732	F_2	..Do	6	.Do	..	..	..	..	..	..	1	2	4	3	..	7.9
11	806	4052	F_2	..Do	5	5257	F_2	..Do	5	.Do	..	..	..	1	1	..	3	..	..	6	1	7.8
12	776	1132	F_2	C×WL	3	2732	P.	C	8	DR×R	..	1	1	..	1	1	3	6	8	..	..	7.6
13	801	6869	$F_{1,2}$	B×F_1	6	4562	F_2	M×B	7	R×R	..	..	..	..	1	1	2	1	5	1	1	7.4
14	814	186	F_1	T×B	4	3975	F_2	B×T	9	DR×R	..	2	1	0	1	3	0	1	3	6	5	7.2
15	814	4683	F_2	..Do	2	3975	F_2	..Do	9	.Do	..	..	..	..	3	2	3	1	1	5	1	7.1
16	767	2104	F_2	WL×B	3	3116	F_1	..Do	9	.Do	..	..	..	1	4	1	2	7	6	1	0	7.1
17	801	2526	F_1	..Do	2	5399	F_2	M×B	8	.Do	..	..	..	..	1	..	..	1	..	..	1	7.0
18	806	3936	F_2	M×B	10	5257	F_2	..Do	9	R×R	..	..	..	1	..	2	..	..	2	..	1	7.0
19	839	5383	F_2	L×M×B	2	4348	F_2	L×M×B	3	DR×DR	..	..	..	1	1	..	..	..	..	1	1	7.0
20	801	5515	F_2	B×T	4	5399	F_2	M×B	8	DR×R	..	..	..	1	1	2	2	..	1	1	3	6.9
21	732	1003	F_2	M×B	9	2442	F_2	..Do	6	R×R	..	..	..	3	7	7	7	7	5	2	..	6.8
22	839	1892	$F_{1,2}$	L×M×B	6	4348	F_2	L×M×B	3	R×DR	..	..	2	1	..	1	..	..	..	2	2	6.8
23	806	4196	F_2	M×B	2	5257	F_2	M×B	9	DR×R	..	..	2	2	2	1	..	..	3	3	..	6.7
24	801	2526	F_1	WL×B	2	5399	F_2	..Do	8	.Do	..	..	..	1	..	..	1	..	..	1	0	6.7
25	801	6861	$F_{1,2}$	B×T	7	4562	F_2	..Do	7	R×R	..	..	..	..	2	1	..	..	..	..	1	6.5
26	767	872	F_2	..Do	5	3116	F_1	B×T	9	DR×R	1	0	0	1	4	6	9	4	4	6	3	6.5
27	801	4263	F_2	..Do	3	4562	F_2	M×B	7	.Do	..	..	..	3	3	1	1	..	4	1	6	6.5
28	767	181	F_1	..Do	4	3116	F_1	B×T	9	.Do	..	..	1	2	6	13	11	5	11	8	3	6.5
29	814	862	F_2	..Do	1	3975	F_2	..Do	9	.Do	..	..	..	2	2	5	2	1	1	3	2	6.3
30	801	872	F_2	..Do	5	5399	F_2	M×B	8	.Do	..	..	..	1	3	8	5	2	..	2	4	6.3
31	839	5389	F_2	M×B	7	4348	F_2	..Do	3	R×DR	..	..	..	6	4	1	0	1	1	1	6	6.2
32	801	872	F_2	B×T	5	4562	F_2	..Do	7	DR×R	..	..	..	1	5	4	3	1	2	2	2	6.1
33	767	190	F_1	..Do	4	3116	F_1	B×T	9	.Do	..	..	..	5	6	11	12	7	4	9	..	6.1
34	801	1892	F_1	M×B	3	4562	F_2	M×B	7	.Do	..	..	1	..	..	..	..	..	2	..	..	6.0
35	801	5515	F_2	B×T	4	4562	F_2	..Do	7	.Do	..	..	1	..	..	1	3	..	1	1	..	6.0
36	731	248	F_2	M×B	4	1249	F_2	WL×B	7	.Do	..	..	..	2	3	3	..	4	6	5	2	6.0
37	732	1228	F_2	..Do	8	2442	F_2	M×B	6	R×R	..	..	..	2	8	5	6	2	8	3	..	6.0
38	732	690	F_1	..Do	5	2442	F_2	..Do	6	DR×R	2	0	6	2	5	5	7	16	10	6	..	6.0
39	751	1919	F_2	WL×B	8	1139	F_2	L×B	5	.Do	..	..	..	5	4	6	6	1	11	1	..	5.9
40	732	618	F_1	M×B	8	2442	F_2	M×B	6	.Do	..	1	2	3	2	5	3	5	9	1	..	5.8
41	731	1245	F_2	WL×B	9	1249	F_2	WL×B	7	R×R	..	..	1	2	1	8	2	3	6	..	..	5.8
42	760	354	F_1	B×T	5	1270	F_2	B×T	2	R×DR	..	..	1	3	9	5	8	4	7	2	..	5.7
43	701	1915	F_2	WL×B	8	1898	F_2	WL×B	3	.Do	..	..	..	7	4	3	2	3	..	1	..	5.7
44	801	6869	$F_{1,2}$	B(M×B)	6	5399	F_2	..Do	8	DR×R	..	..	..	..	1	..	..	..	..	1	..	5.7
45	801	4570	F_2	B×T	2	4562	F_2	..Do	7	.Do	..	..	1	2	5	3	1	1	1	4	..	5.6
46	814	703	F_1	..Do	4	3975	F_2	B×T	9	.Do	..	..	..	3	5	2	7	5	6	2	4	5.5
47	732	953	F_2	M×B	3	2442	F_2	M×B	6	.Do	..	2	2	3	8	9	5	3	7	6	..	5.5
48	801	7528	F_1	..Do	4	4562	F_2	..Do	8	.Do	..	..	..	1	4	2	1	..	1	..	1	5.3
49	731	2116	F_2	..Do	10	1249	F_2	WL×B	7	R×R	..	1	1	2	3	0	2	2	2	1	..	5.2
50	745	2115	F_2	C×T	4	1258	F_2	B×T	4	DR×DR	..	..	..	2	1	6	4	2	..	..	..	5.2
51	801	6843	F_2	B×T	3	4562	F_2	..Do	8	DR×R	..	..	..	3	2	1	..	2	1	..	..	5.1
52	801	2831	F_1	M×B	4	5399	F_2	..Do	8	.Do	1	1	1	4	1	7	2	2	2	..	2	5.0
53	801	1892	F_1	..Do	3	5399	F_2	..Do	8	.Do	1	1	..	1	2	..	1	0	1	1	1	5.0
54	801	7528	F_1	..Do	4	4562	F_2	..Do	8	.Do	..	..	..	1	2	1	2	..	1	1	..	5.0
55	731	1755	F_2	WL×B	6	1249	F_2	WL×B	7	R×R	..	..	..	2	1	4	1	2	..	..	..	5.0
56	745	2513	F_2	C×T	4	1258	F_2	B×T	4	DR×DR	..	..	..	..	2	5	2	..	3	1	..	5.0
57	839	3950	F_2	M×B	4	4348	F_2	M×B	3	.Do	..	2	3	3	2	4	1	1	1	2	2	4.95
58	754	873	F_2	B×T	3	871	F_2	B×T	2	.Do	1	2	4	1	..	..	..	1	8	..	..	4.94
59	806	599	F_2	M×B	3	5257	F_2	M×B	7	DR×R	..	..	2	1	2	2	1	..	1	..	..	4.86
60	760	300	F_1	B×T	7	1270	F_2	B×T	2	R×DR	..	..	2	19	8	13	6	4	5	2	1	4.83
61	806	4456	F_2	M×B	1	5257	F_2	M×B	7	DR×R	..	1	1	1	1	..	1	..	1	1	..	4.71

BOOTING.

TABLE 38.—*Distribution of boot-grades in offspring of parents one or both of which belong to a hybrid generation beyond the first*—Continued.

B = Brahma; C = Cochin; G = Game; L = Leghorn; M = Minorca; S = Silkie; Sp = Spanish; T = Tosa; WL = White Leghorn.

Serial No.	Pen No.	Mother No.	Gen.	Races	Gr.	Father No.	Gen.	Races	Gr.	Mating	0	1	2	3	4	5	6	7	8	9	10	Av.
62	732	2407	F₂	M×B	2	2442	F₂	M×B	6	DR×R	..	..	1	3	4	1	1	2	..	1	..	4.69
63	701	894	F₂	L×B	7	1898	F₂	L×B	3	R×DR	1	1	2	8	6	1	2	2	4	2	..	4.62
64	760	994	F₂	B×T	3	1270	F₂	B×T	3	DR×DR	..	..	..	..	4	2	1	..	..	..	..	4.57
65	760	981	F₂	..Do	3	1270	F₂	..Do	3	..Do	1	..	3	6	1	4	2	7	..	..	..	4.54
66	701	1772	F₂	L×B	6	1898	F₂	L×B	3	R×DR	..	..	..	4	7	2	2	2	..	..	..	4.47
67	839	3541	F₁	M×B	6	4348	F₂	M×B	3	DR×DR	4	1	4	..	..	4	2	..	2	1	2	4.30
68	842	1645	F₂	..Do	2	4385	F₂	..Do	4	..Do	3	2	6	5	6	6	3	0	2	4	1	4.29
69	770	2049	F₁	L×B	3	926	F₂	..Do	3	..Do	9	3	1	6	8	2	6	6	3	1	3	4.29
70	731	2577	F₁.₂	L×C	4	1249	F₂	L×B	7	DR×R	..	..	2	2	3	2	2	1	..	..	..	4.25
71	701	250	F₁	L×B	3	1898	F₂	..Do	3	DR×DR	3	3	5	8	12	10	10	6	1	..	..	4.22
72	701	1335	F₂	T×L×B	8	1898	F₂	..Do	3	R×DR	..	..	1	9	6	6	4	1	..	..	..	4.22
73	806	4767	F₂	M×B	3	5257	F₂	M×B	7	DR×R	..	..	1	..	2	1	1	..	..	..	..	4.20
74	740	1439	F₂	C×L	2	1145	F₂	C×L	3	DR×DR	3	..	1	3	6	4	2	..	2	1	..	4.18
75	754	3126	F₂	B×T	4	871	F₂	B×T	3	..Do	..	2	5	11	7	10	5	0	2	1	..	4.14
76	770	1645	F₂	M×B	4	926	F₂	M×B	3	..Do	..	1	9	5	5	2	2	3	1	..	..	4.10
77	731	249	F₁	L×B	3	1249	F₂	L×B	7	DR×R	7	4	6	5	7	5	9	3	6	1	..	4.08
78	732	703	F₁	M×B	3	2442	F₂	M×B	6	..Do	1	3	13	13	8	6	7	6	3	..	..	4.07
79	770	720	F₁	B×L	4	926	F₂	..Do	3	DR×DR	6	1	3	9	5	4	5	1	4	1	1	4.05
80	732	2441	F₂	M×B	0	2442	F₂	..Do	6	DR×R	..	1	6	8	2	6	0	3	1	..	..	4.00
81	760	1042	F₂	B×T	3	1270	F₂	B×T	2	DR×DR	2	3	3	9	3	5	3	2	0	1	..	4.00
82	731	384	F₁	L×B	4	1249	F₂	L×B	7	DR×R	2	1	4	4	3	2	4	0	1	1	..	3.82
83	814	4566	F₂	B×T	2	3975	F₂	B×T	9	..Do	1	4	2	4	3	2	1	..	..	..	..	3.82
84	732	599	F₁	M×B	3	2442	F₂	M×B	6	..Do	6	5	23	10	5	3	4	5	8	3	1	3.78
85	770	761	F₁	B×L	3	926	F₂	..Do	3	DR×DR	7	3	5	3	7	7	2	6	1	1	..	3.71
86	731	1770	F₂	..Do	7	1249	F₂	L×B	7	..Do	1	..	8	6	9	3	2	..	2	..	..	3.65
87	861	5165	F₂	T×C	10	95	F₁	T×C	5	R×DR	..	..	..	10	3	2	1	..	..	..	..	3.63
88	754	3175	F₂	B×T	2	871	F₂	B×T	2	DR×DR	1	..	2	1	3	4	..	..	..	..	..	3.55
89	731	2102	F₂	L×B	1	1249	F₂	L×B	7	DR×R	1	0	4	2	4	1	..	1	..	..	..	3.43
90	840	1755	F₂	M×B	6	4177	F₂	..Do	2	R×DR	..	..	6	7	7	3	1	..	..	..	..	3.42
91	701	2576	F₂	L×B	2	1898	F₂	..Do	3	DR×R	2	1	1	8	11	2	1	..	..	..	..	3.35
92	842	2049	F₁	..Do	3	4385	F₂	M×B	4	..Do	11	1	2	8	5	3	1	0	2	2	2	3.35
93	754	853	F₁	B×T	1	871	F₂	B×T	3	..Do	2	3	4	6	4	1	6	..	..	..	..	3.31
94	826	2652	F₂	M×B	3	4093	F₂	M×B	0	..Do	8	2	1	8	1	..	..	1	2	3	..	3.28
95	754	1052	F₂	B×T	2	871	F₂	B×T	2	..Do	3	..	7	9	9	5	2	..	..	..	..	3.26
96	701	965	F₂	T×L×B	0	1898	F₂	L×B	3	..Do	1	4	6	12	8	4	0	2	0	..	..	3.19
97	732	1833	F₂	M×B	1	2442	F₂	M×B	6	DR×R	1	1	7	6	6	4	1	..	..	..	..	3.19
98	732	631	F₁	..Do	3	2442	F₂	..Do	6	..Do	3	4	10	16	12	4	1	2	..	..	..	3.06
99	754	862	F₁	B×T	1	871	F₂	B×T	2	DR×DR	1	5	10	17	10	4	1	..	..	..	..	2.96
100	837	5641	F₂	T×L×B	0	4288	F₂	L×B	2	..Do	1	2	2	3	..	2	..	1	..	..	..	2.91
101	840	3841	F₂	L×B	0	4177	F₂	..Do	2	D×DR	3	3	2	6	4	2	..	..	..	..	..	2.86
102	701	721	F₁	..Do	2	1898	F₂	..Do	3	DR×DR	2	4	3	8	3	2	2	..	..	..	..	2.83
103	839	3949	F₂	..Do	4	4348	F₂	..Do	3	..Do	1	2	..	..	1	1	..	..	..	..	..	2.83
104	840	732	F₁	..Do	3	4177	F₂	..Do	2	..Do	7	6	9	8	7	2	1	2	..	1	..	2.67
105	840	249	F₁	..Do	3	4177	F₂	..Do	2	..Do	7	3	5	6	2	9	..	..	..	..	..	2.62
106	840	3916	F₁.₂	..Do	2	4177	F₂	..Do	2	..Do	5	1	4	2	2	2	1	..	..	..	..	2.29
107	842	4945	F₂	M,L×B	1	4385	F₂	M×B	4	..Do	9	3	6	5	1	2	4	..	..	..	..	2.27
108	731	2595	F₁	L×B	1	1249	F₂	L×B	7	D×R	6	6	7	1	..	..	..	..	..	..	..	2.15
109	840	5169	F₂	..Do	3	4177	F₂	..Do	2	DR×DR	6	2	5	5	2	2	..	..	..	..	..	2.05
110	837	5667	F₂	..Do	2	4288	F₂	..Do	2	..Do	2	1	2	1	1	..	..	..	..	..	..	2.00
111	749	1355	F₂	G×C	2	1854	F₂	G(C×L)	0	DR×D	..	2	5	1	..	..	..	..	..	..	..	1.87
112	824	3901	F₂	M×S	1	5095	F₂	M×S	1	DR×DR	17	3	2	3	..	2	..	..	1	2	..	1.73
113	751	1254	F₂	L×B	0	1139	F₂	L×B	8	D×R	17	5	5	4	3	3	..	1	..	..	..	1.63
114	749	816	F₁	..Do	2	1854	F₂	G(C×L)	0	DR×D	6	7	3	5	1	..	..	..	..	..	..	1.45
115	749	929	F₁	G×C	0	1854	F₂	..Do	0	D×D	8	3	1	..	..	..	1	..	..	1	..	1.43
116	749	819	F₁	L×B	1	1854	F₁.₂	G(C×L)	0	DR×D	9	3	5	3	..	..	..	..	..	..	..	1.10
117	804	5099	F₂	S×Sp	0	3823	F₁	S×Sp	0	D×D	2	..	..	1	..	..	..	..	..	..	..	1.00
118	804	6043	F₂	..Do	1	3823	F₁	..Do	0	..Do	..	1	..	..	..	..	..	..	..	..	..	1.00
119	817	5730	F₂	L×Sp	0	3900	F₂	..Do	1	D×DR	3	3	1	..	..	..	..	..	..	..	..	0.71
120	817	4696	F₂	..Do	0	3900	F₂	..Do	1	..Do	9	7	3	..	..	..	..	..	..	..	..	0.68
121	817	6046	F₂	S×M	0	3900	F₂	..Do	1	..Do	10	..	3	..	..	..	..	..	..	..	..	0.46
122	817	6833	F₁.₂	L(G×S)	0	3900	F₂	..Do	1	..Do	6	2	1	..	..	..	..	..	..	..	..	0.44
123	817	5062	F₁	L(Sp)	0	3900	F₂	..Do	1	..Do	18	7	2	..	..	..	..	..	..	..	..	0.41
124	817	5069	F₁	..Do	0	3900	F₂	..Do	1	..Do	21	7	2	..	..	..	..	..	..	..	..	0.37
125	817	6406	F₁	..Do	0	3900	F₂	..Do	1	..Do	25	8	2	..	..	..	..	..	..	..	..	0.34
126	817	7047	F₁	..Do	0	3900	F₂	..Do	1	..Do	4	2	..	..	..	..	..	..	..	..	..	0.33
127	749	2651	F₂	G×C	0	1854	F₂	G(C×L)	0	D×D	2	1	..	..	..	..	..	..	..	..	..	0.33
128	824	4714	F₂	S×Sp	0	5095	F₂	M×S	1	..Do	26	6	1	1	..	..	..	..	..	..	..	0.32
129	817	4690	F₁	..Do	0	3900	F₂	S×Sp	1	D×DR	21	6	1	..	..	..	..	..	..	..	..	0.29
130	824	7439	F₂	..Do	0	5095	F₂	M×S	1	D×D	11	4	..	..	..	..	..	..	..	..	..	0.27
131	804	4715	F₂	..Do	0	3823	F₁	S×Sp	0	DR×DR	18	2	1	..	..	..	..	..	..	..	..	0.19
132	804	3898	F₂	S×M	0	3823	F₁	..Do	0	D×DR	19	..	..	..	..	..	..	..	..	..	..	0.00
133	804	3902	F₂	..Do	0	3823	F₁	..Do	0	..Do	33	..	..	..	..	..	..	..	..	..	..	0.00
134	804	4657	F₂	..Do	0	3823	F₁	..Do	0	..Do	8	..	..	..	..	..	..	..	..	..	..	0.00
135	804	4716	F₂	..Do	0	3823	F₁	..Do	0	..Do	19	..	..	..	..	..	..	..	..	..	..	0.00
136	804	5431	F₂	..Do	0	3823	F₁	..Do	0	..Do	16	..	..	..	..	..	..	..	..	..	..	0.00

In table 38 I have given in the section lying between that headed "Father" and that headed "Offspring" the "Matings." This column differs from the others of the table in not being, in general, based upon observation, but upon a sometimes complicated judgment. Of course, all of the F_1 generation, where this generation occurs, may be taken as of DR composition; but the decision as to whether a given individual of F_2 is a DR, an extracted dominant, or an extracted recessive is not always easy, because of the manifestation of imperfect dominance. But the assignments are by no means arbitrary. Taking the Brahma crosses, which are by far the most numerous, we see, from tables 31, B and 33, that those F_2 individuals that have a boot of grade 6 or higher are almost certainly extracted recessives (which are equivalent to pure-bred Dark Brahmas). Those with a grade of 3 or even 4 and lower to 2 or even 1 are probably heterozygotes, while those with grade 0 and some of those with grade 1 are extracted dominants. In cases of doubt the distribution of grades in the offspring will give the deciding vote. In case the individual has been used as a parent in more than one mating the results in all the matings are taken into account, for the germinal constitution of an individual must be regarded as fixed at all times and in all matings. The assignment under "Matings" has, then, been made by the application of the above rules.

In tables 39 to 43 there are grouped together the progeny from matings of the same sort, selecting from table 38 the crosses into which the Dark Brahma enters as the booted parent.

TABLE 39.—*RR × RR crosses from table 38.*

Serial No.	Boot-grade in offspring.												Parental grades.		
	0	1	2	3	4	5	6	7	8	9	10	Avge.	Female.	Male.	Average.
1	..	..	..	..	..	..	..	..	..	10	15	9.6	7	9	8.0
2	..	..	..	..	..	..	..	..	..	1	1	9.5	4	8	6.0
3	..	..	..	..	..	..	..	..	1	3	4	9.4	5	9	7.0
4	..	..	..	..	..	..	..	..	1	1	2	9.3	6	7	6.5
5	..	..	..	..	..	1	..	..	3	4	9	9.1	9	9	9.0
6	..	..	..	..	..	..	1	1	1	..	2	7	8.8	8	9	8.5
7	..	..	..	..	..	..	..	1	1	1	4	8.8	2	9	5.5
8	..	..	..	..	..	..	..	1	2	2	3	8.6	7	5	6.0
9	..	..	..	..	1	..	..	1	4	2	3	8.3	7	5	6.0
10	..	..	..	..	..	..	1	2	4	3	..	7.9	8	6	7.0
11	..	..	..	..	1	1	..	3	..	6	1	7.8	5	5	5.0
13	..	..	..	..	..	1	1	2	..	1	1	7.4	6	7	6.5
18	..	..	..	..	1	..	2	..	2	..	1	7.0	10	9	9.5
21	..	..	..	..	3	7	7	7	7	5	2	6.8	9	6	7.5
25	..	..	..	..	..	2	1	..	..	..	1	6.5	7	7	7.0
37	..	..	..	2	8	5	6	2	8	3	..	6.0	8	6	7.0
39	..	..	..	5	4	6	6	1	11	1	..	5.9	8	8	8.0
41	..	..	1	2	1	8	2	3	6	..	..	5.8	9	7	8.0
49	..	1	1	1	2	3	..	2	2	1	..	5.2	10	7	8.5
55	..	..	..	2	1	4	1	2	..	..	..	5.0	6	7	6.5
Totals (287).	..	1	2	12	22	39	30	28	53	46	54	7.25			
Per cent....	..	0.3	0.7	4.2	7.7	13.6	10.5	9.8	18.5	16.0	18.8	...			

The significance of the data given in tables 39 to 43 is best brought out by summarizing them. Especially instructive is a comparison of the pure-bred with the hybrids. Since the data are most complete in the case of the Brahma crosses, these will be considered in most detail. So far as they go, the results with the Cochins and Silkies are entirely confirmatory.

TABLE 40.—DR × RR crosses from table 38.

Serial No.	Boot-grade in offspring.											Average.
	0	1	2	3	4	5	6	7	8	9	10	
14	..	2	1	..	1	3	..	1	3	6	5	7.2
15	..	..	..	..	3	2	3	1	1	1	5	7.1
16	..	..	..	1	4	1	2	7	6	1	..	7.1
17	..	..	..	..	1	..	..	1	..	..	1	7.0
20	..	..	..	1	1	2	2	..	1	1	3	6.9
22	..	..	..	2	1	..	1	..	..	2	2	6.8
23	..	..	..	2	2	2	1	..	..	3	3	6.7
24	..	..	..	..	1	..	..	1	..	1	..	6.7
26	1	..	..	1	4	6	9	4	4	6	3	6.5
27	..	..	..	..	2	3	..	3	1	4	1	6.5
28	..	..	1	2	6	13	11	5	11	8	3	6.3
29	..	..	..	2	2	5	2	1	1	3	2	6.3
30	..	..	..	1	3	8	5	2	..	2	4	6.3
31	..	..	..	6	4	1	..	1	1	1	6	6.2
32	..	..	..	1	5	4	3	1	2	2	2	6.1
33	..	..	..	5	6	11	12	7	4	9	..	6.1
34	..	..	1	..	..	..	1	..	2	..	..	6.0
35	..	..	1	..	..	1	3	..	1	1	..	6.0
36	..	..	2	3	3	..	2	..	..	5	2	6.0
40	..	1	2	3	2	5	3	5	9	1	..	5.8
42	..	..	1	3	9	5	8	4	7	2	..	5.7
43	..	..	..	..	7	4	3	2	3	1	..	5.7
44	..	..	..	1	..	1	..	..	..	1	..	5.7
45	..	..	1	2	5	3	1	1	1	4	..	5.6
46	..	..	3	5	2	7	5	6	2	4	..	5.5
47	..	2	2	3	8	9	5	3	7	6	..	5.5
48	..	..	..	1	4	2	1	..	1	..	1	5.3
51	..	..	..	..	3	2	1	..	1	..	..	5.1
52	1	1	1	4	1	7	2	2	2	..	2	5.0
53	1	1	..	1	2	..	1	..	1	1	1	5.0
54	..	..	..	1	2	1	1	..	1	..	..	5.0
59	..	..	2	1	2	2	1	..	1	..	..	4.9
60	..	..	2	19	8	13	6	4	5	2	1	4.8
61	..	1	1	1	1	..	1	..	1	1	..	4.8
62	..	..	1	3	4	1	1	2	..	1	..	4.7
63	1	1	2	8	6	1	2	2	4	2	..	4.6
66	..	..	..	4	7	2	2	2	..	..	..	4.5
70	..	..	2	2	3	2	2	1	..	..	..	4.3
72	..	..	1	9	6	6	4	1	..	..	..	4.2
73	..	..	1	..	2	1	1	..	..	..	..	4.2
77	7	4	6	5	7	5	9	3	6	1	..	4.1
78	1	3	13	13	8	6	7	6	8	..	..	4.1
80	..	1	6	8	2	6	..	3	1	..	..	4.0
82	2	1	4	4	3	2	4	..	1	1	..	3.8
83	1	4	2	4	3	2	1	..	..	..	..	3.8
84	6	5	23	10	5	3	4	5	8	3	1	3.8
86	1	..	8	6	9	3	2	..	2	..	..	3.7
89	1	..	4	2	4	1	..	1	..	..	..	3.4
90	..	..	6	7	7	3	1	..	..	..	..	3.4
97	1	1	7	6	6	4	1	..	..	..	..	3.2
98	3	4	10	16	12	4	1	2	..	..	..	3.1
Total (1199)	27	32	117	181	200	172	142	88	105	87	48	5.04
Per cent...	2.3	2.7	9.8	15.1	16.7	14.3	11.9	7.3	8.8	7.2	4.0	...

TABLE 41.—DR × DD crosses.

Serial No.	Boot-grade in offspring.						Average.
	0	1	2	3	4	5	
101	3	3	2	6	4	2	2.9
113	6	7	3	5	1	...	1.5
116	9	3	5	3	...	...	1.1
Total (62)	18	13	10	14	5	2	1.69
Per cent	29.5	21.3	16.4	23.0	8.2	1.6	...

TABLE 42.—DR × DR crosses.

Serial No.	Boot-grade in offspring.											Average.
	0	1	2	3	4	5	6	7	8	9	10	
19	..	..	..	..	1	1	..	..	..	1	..	7.0
54	..	..	..	2	1	6	4	2	..	..	..	5.2
56	..	..	..	2	5	2	..	3	1	..	..	5.0
57	..	2	3	3	2	4	1	1	1	2	2	5.0
58	1	2	1	4	1	..	..	..	8	..	..	4.9
59	..	..	2	1	2	2	1	..	1	..	..	4.9
64	..	..	..	..	4	2	1	..	..	..	..	4.6
65	1	..	3	6	1	4	2	7	..	..	2	4.5
67	4	1	4	..	..	4	2	..	2	1	2	4.3
68	3	2	6	5	6	6	3	..	2	4	1	4.3
69	9	3	1	6	8	2	6	6	3	1	3	4.3
71	..	..	2	2	3	2	2	1	..	..	..	4.3
75	..	2	5	11	7	10	5	..	2	1	..	4.1
76	3	2	1	9	5	5	2	2	3	1	..	4.1
79	6	1	3	9	5	4	5	1	4	1	1	4.1
81	2	3	3	9	3	5	8	2	..	1	..	4.0
85	7	3	5	3	7	7	2	6	1	1	..	3.7
88	1	..	2	1	3	4	..	..	..	..	..	3.6
91	2	1	1	8	11	2	1	..	..	..	..	3.4
92	11	1	2	8	5	3	1	..	2	2	2	3.4
93	2	3	4	6	4	1	6	..	..	..	..	3.3
94	8	2	1	8	1	..	..	1	2	3	..	3.3
95	3	..	7	9	5	2	..	..	..	..	..	3.3
96	1	4	6	12	8	4	..	2	..	..	..	3.2
99	1	5	10	17	10	4	1	..	..	..	..	3.0
100	1	2	2	3	..	2	..	1	..	..	..	2.9
102	2	4	3	8	3	2	2	..	..	..	..	2.8
103	1	2	..	..	1	1	1	..	..	..	..	2.8
104	7	6	9	8	7	2	1	2	..	1	..	2.7
105	7	3	5	6	2	9	..	..	..	..	..	2.6
106	5	1	4	2	2	2	1	..	..	..	..	2.3
107	9	3	6	5	1	2	4	..	..	..	..	2.3
109	6	2	5	5	2	2	..	..	..	..	..	2.1
110	2	1	2	..	1	1	..	..	..	..	..	2.0
Total (851)	105	61	108	178	127	109	62	37	32	20	12	3.59
Per cent...	12.3	7.2	12.7	20.9	14.9	12.8	7.3	4.4	3.8	2.3	1.4	...

TABLE 43.—DD × DD (Silkie crosses).

Serial No.	Boot-grade in offspring.				Average.
	0	1	2	3	
117	2	..	..	1	1.00
118	..	1	..	..	1.00
128	26	6	1	1	0.32
130	11	4	..	..	0.27
131	18	2	1	..	0.19
132	19	..	..	..	0.0
133	33	..	..	..	0.0
134	8	..	..	..	0.0
135	19	..	..	..	0.0
136	16	..	..	..	0.0
Total (169)	152	13	2	2	0.14
Per cent	89.9	7.7	1.2	1.2	....

Table 44 shows clearly, first, that there are families of two booted parents that never fail to produce booted offspring. There is, however, even in pure-bred booted races, a marked variability in the grade of booting, extending from 3 (or 4) to 10. The significance of this variability must be left for future investigations. There is in the least boot, as it were, an extension of the field of activity of the feather-inhibiting factor that is always present on the hinder aspect of the shank, so that it interferes with the development of feathers on the inner face of the shank also.

In the first hybrid generation all somatic cells are hybrid. The feather inhibitor is present in the skin of the shank, but its strength is diluted by the presence in the same cells of a protoplasm devoid of the inhibiting property. Consequently, the prevailing grade of the boot falls from 6 (or 10) to 3. Despite the dilution, inhibition is complete in about 8 per cent of the offspring (grade 0); in about 10 per cent of the offspring the inhibiting factor is so weak that the boot develops as in the pure-blooded Brahma. When, as a result of inbreeding F_1's, the feather-inhibiting factor is eliminated from certain offspring, and such full-feathered birds are bred together, we find a return of the mode to high numbers, such as 8 to 10 (but also 5). There is no doubt of segregation.

TABLE 44.—*Brahma crosses.* (*All entries are percentages.*)

Percentage.	From table.	Boot-grade in offspring.											Average grade.
		0	1	2	3	4	5	6	7	8	9	10	
Pure blood	31, B	....	....	....	3.3	3.3	6.6	24.6	4.9	9.8	14.8	32.8	7.62
F_1 (D×R)	32	7.9	13.8	16.8	31.0	17.5	7.8	3.4	1.1	0.7	....	....	2.84
Extracted R×R	39	....	0.3	0.7	4.2	7.7	13.6	10.5	9.8	18.5	16.0	18.8	7.25
DR×RR	40	2.3	2.7	9.8	15.1	16.7	14.3	11.9	7.3	8.8	7.2	4.0	5.04
		50 p. ct. DR.						50 p. ct. RR.					
DR×DR	42	12.3	7.2	12.7	20.9	14.9	12.8	7.3	4.4	3.8	2.3	1.4	3.59
		25 p. ct. DD.		50 p. ct. DR.				25 p. ct. RR.					
DR×DD	41	29.5	21.3	16.4	23.0	8.2	1.6	....	....	....	....	....	1.69
		50 p.ct. DD.		50 p. ct. DR.									

If a heterozygous bird be mated to a recessive the variability of the offspring is much increased, owing to the occurrence in the progeny of both DR and RR individuals (table 40). The offspring do not, to be sure, fall into two distinct and well-defined types, as in typical Mendelian cases; but one part of the range of variation agrees fairly with that of pure RR's, *i. e.*, Brahmas, and the remainder with that of heterozygotes. And if we make the division in the middle of the middle class, viz, 5, we shall find a close approximation to that equality of extracted recessives and heterozygotes that the segregation theory calls for (table 44).

If, again, two heterozygous birds be mated, the variability is still greater and the proportion of clean-footed offspring rises to 12 per cent. These, together with some of the extremely slightly booted offspring, represent the extracted dominants. The whole range now falls into three regions divided by the middle of grades 2 and 5. These regions correspond to the DD's, the DR's, and the RR's of typical cases of segregation, and their relative proportions are approximately as 25 : 50 : 25.

Finally, if a heterozygote be mated to an extracted dominant the proportion of clean-footed offspring rises to about 30 per cent and the whole range of variation falls readily into two parts, the one comprising grades 0 and 1, the other grades 2 and above. The first includes the DD offspring;

the second, the DR's; and their frequency is equal. One will not fail to note that we are not here dealing with a case of blending simply, and the inheritance of the blend; such a view is negatived by the fact of the much greater variability of DR × DR cross over the simple D × R cross of the first generation. One may safely conclude, then, that, despite the apparent blending of booting characters in the first generation of hybrids, true segregation takes place. But this is always to be seen through the veil of imperfect dominance.

A casual examination of table 38 would seem to show a correlation between the grade of booting of the parents and that of the average of their progeny. Thus, on the whole, the parental grades run high in the upper part of the table and run low in the lower part. This relation would thus seem to confirm Castle's conclusion for polydactylism in guinea-pigs that there is an inheritance of the degree of a character. One consequence of such an inheritance would be that it would be possible in a few generations to increase or diminish the grade of a character and fix any required grade in the germ-plasm. A more careful consideration of the facts of the case shows that this relation has another interpretation. The grade of boot of the different parents varies largely because their gametic constitution is diverse. As table 39 shows, the parents of the upper part of table 38 are chiefly extracted recessives, and consequently their booting and that of their offspring are characterized by high grades. On the other hand, the parents of the lower part of the table are heterozygous or extracted dominants and, consequently, their grades and also those of their offspring average low. On account of the lack of homogeneity of the families in table 38, one can draw from it no proper conclusions as to relation between parental and filial grades. On the other hand, from a homogeneous table, like table 39, we can hope to reach a conclusion as to the existence of such a relation. I have calculated, in the usual biometric fashion, the coefficient of correlation between average parental and filial grades, and found it to be -0.17 ± 0.13. This can only be interpreted to mean that in a homogeneous assemblage of families there is no correlation between the grade of booting of parents and offspring.

CHAPTER VII.

NOSTRIL-FORM.

In my 1906 report I described in detail the form of the nostril in poultry. Usually it is closed down to a narrow slit, but in some races, as, e. g., the Polish and Houdans, the closing flap of skin fails to develop and the nostril remains wide open. This is apparently an embryonic condition. Thus in Keibel and Abraham's (1900) Normaltafeln of the fowl it is stated that the outer nasal opening, which is at first wide open, becomes closed with epithelium at about the middle of the sixth day of development. The Polish and Houdan fowl thus retain in the outer nasal opening an embryonic condition. The question is: How does this embryonic, open condition of the nostril behave in heredity with reference to the more advanced narrow-slit condition?

The wide-nostriled races used were both the Polish and the Houdan. The condition of the external nares is much the same in the two, but is slightly more exaggerated in the Houdans than in the Polish. The open nostril is often associated with a fold across the culmen, apparently due to the upturning of the anterior end of the premaxillary process of the nasal bone. Breeders of Houdans have sought to exaggerate the height of the fold. In both races there is great variability in the degree of "openness" of the nostril, and to indicate this I have adopted a scale of 10 grades (running from 1, the narrowest, to 10, the widest). To get some idea of this variability let us consider the grade of nostril in some families of pure Houdans.

TABLE 45.—*Variability (expressed in decimal grades) of the degree of "openness" of the nostrils in families of "pure-bred" Houdans.*

Serial No.	Pan No.	Mother.		Father.		Grade of openness in offspring.									
		No.	Grade.	No.	Grade.	1	2	3	4	5	6	7	8	9	10
1	727	2457	9	831	10	..	..	..	..	..	..	..	..	5	4
2	727	2459	10	831	10	..	..	1	..	..	..	1	3	7	3
3	727	2494	9	831	10	..	..	..	..	..	..	..	..	1	4
4	727	3105	9	831	10	..	1	..	1	2	1	..	5	7	3
5	727	3106	9	831	10	..	..	..	..	..	..	..	..	2	1
6	803	2457	8	7522	9	..	1	1	..	..	2	4	7	10	3
7	803	2459	10	7522	9	..	..	..	..	..	..	1	6	4	2
8	803	3105	9	7522	9	1	..	..	..	4	2	2	7	3	7
Totals (119)						1	2	2	1	6	5	8	28	39	27
Percentages							5.3			5.3	4.4	7.1	24.8	34.5	23.9

Table 45 shows that the prevailing grade in the offspring of pure Houdans is 9; that grades 8 and 10 are also extremely common; and that lower grades, even down to 1, may occur, but these are much less common.

We have next to consider the grade-distribution of the offspring of the narrow mated with the wide nostril.

TABLE 46.—*Distribution of the frequency of the different grades of "openness" of nostril when one parent has the open nostril and the other the closed.*

Serial No.	Pen No.	Mother.				Father.				Grade of openness in offspring.									
		No.	Gen.	Race.	Gr.	No.	Gen.	Races.	Gr.	1	2	3	4	5	6	7	8	9	10
9	727	121	P.	Dk. Brahma.	1	831	P.	Houdan	10	9	11	6	6	2	3	1	1	..	..
10	735	142	P.	Mediterran..	1	30	P.	Polish	8	4	1	..	..	..	..	..	..	..	..
11	735	177	P.	....Do......	1	30	P.	...Do..	8	..	4	2	1	..	..	..	..	..	..
12	735	198	P.	....Do......	1	30	P.	....Do..	8	..	3	1	..	..	1	..	..	..	..
				Totals (56)						13	19	9	7	2	4	1	1	..	..
				Percentages						23.2	34.0	16.1	12.5	3.6	7.1	1.8	1.8	..	..
*12a	813	912	F₂	Houd × Legh.	2	3904	F₂	Houd × Legh.	7	3	10	3	1	1	..	..	..	..	..

* Extracted D×R.

Table 46 gives us a picture of the nature of the dominance in this case. At first sight the narrow nostril, grades 1 and 2, including 57 per cent of the offspring, appears to be dominant. But, as later evidence shows, it is recessive. The wide nostril is dominant, but so imperfectly that only 10 per cent have a nostril above one-half open.

Let us now consider the distribution of nostril form in families whose parents are hybrids of the first or later generation, crossed respectively on recessives, heterozygotes, and dominants (tables 47–49).

TABLE 47.—*Distribution of frequency of the different grades of "openness" of nostril when one parent is heterozygous and the other recessive, i. e., with closed nostril (DR×R).*

Serial No.	Pen No.	Mother.				Father.				Total gr.	Grade of openness in offspring.									
		No.	Gen.	Races.	Gr.	No.	Gen.	Race.	Gr.		1	2	3	4	5	6	7	8	9	10
13	768	298	F₂	Med. × Polish.	2	1689	P.	Med..	1	3	11	9	3	1	3	..	1	..	..	..
14	768	509	F₁	....Do......	1	1689	P.	...Do..	1	2	12	5	6	1	1	..	..	..	..	..
				Totals (53)							23	14	9	2	4	0	1	..	..	..
				Percentages							43.4	26.4	17.0	3.8	7.6	..	1.9	..	..	..

The study of the tables 45 to 54 establishes the following conclusions:

First, high nostril is dominant. This means that there is a factor that inhibits the development of the narial flap. In the absence of such a factor the flap goes on developing normally. This hypothesis is opposed to the conclusion that I reached in my report of 1906 (pp. 68, 69). I there said:

A close agreement exists between the percentage obtained in each generation and the expectation of the Mendelian theory, assuming that narrow nostril is dominant. The statistics do not, however, tell the whole story. In 36 per cent of the cases in the F_1 generation the nostril was wider than in the "narrow" ancestor. Even in the F_2 generation nearly half of the "narrow and intermediate" were of the intermediate sort. This intermediate form is evidence that dominance is imperfect and segregation is incomplete.

NOSTRIL-FORM.

TABLE 48.—*Distribution of frequency of grades of "openness" in offspring when both parents are heterozygous ($DR \times DR$).*

Serial No.	Pen No.	Mother				Father				Total gr.	Grade of openness in offspring									
		No.	Gen.	Races.	Gr.	No.	Gen.	Races.	Gr.		1	2	3	4	5	6	7	8	9	10
15	802	5314	F_1	Polish × Min..	3	6652	F_1	Polish × Min..	4	7	1	5	5	..	1	..	..	..	3	1
16	805	5307	F_1	...Do..	5	4799	F_1	...Do....	2	7	7	7	13	3	7	1	..	2	2	1
17	852	5104	F_1	Hou.× Dk.Br.	3	5969	F_1	Hou.× Dk.Br.	3	6	4	11	4	2	1	1	1	..	..	..
18	805	4800	F_1	Polish × Min..	3	4799	F_1	Polish × Min..	2	5	10	13	9	1	2	8	..	1	2	..
19	805	5308	F_1	...Do....	3	4799	F_1	...Do....	2	5	5	12	4	..	3	..	1	1	..	..
20	805	5929	F_1	...Do....	3	4799	F_1	...Do....	2	5	3	7	3	2	1	..	..	..	..	..
21	759	797	F_1	Houd.× Min.	3	570	F_1	Houd.× Min.	2	5	2	4	2	2	..	..	..	..	..	2
22	759	797	F_1	...Do....	3	352	F_1	...Do....	1	4	..	2	2	..	..	..	..	..	1	1
23	805	4447	F_1	Polish × Min.	2	4799	F_1	Polish × Min.	2	4	6	5	4	..	2	..	1	1	3	..
24	805	4765	F_1	...Do....	2	4799	F_1	...Do....	2	4	5	12	4	2	1	1	2	..	2	..
25	805	4797	F_1	...Do....	2	4799	F_1	...Do....	2	4	4	2	6	..	..	..	1	..	..	..
26	805	5163	F_1	...Do....	2	4799	F_1	...Do....	2	4	7	17	13	4	1	2	2	2	1	..
27	805	5304	F_1	...Do....	2	4799	F_1	...Do....	2	4	5	9	8	..	1	..	..	..	..	..
28	852	7070	F_1	Hou.× Dk.Br.	1	5969	F_1	Hou.× Dk.Br.	3	4	4	11	4	2	1	1	1	..	..	1
29	759	529	F_1	Houd.× Min.	2	570	F_1	Houd.× Min.	2	4	2	3	..	..	..	..	..	..	..	..
30	759	529	F_1	...Do....	2	352	F_1	...Do....	1	4	1	3	..	..	..	..	..	..	..	..
31	728	174	F_1	Hou.× Wh.L.	1	258	F_1	Hou.× Wh.L.	2	3	2	7	2	1	1	1	..	..	..	..
32	805	4798	F_1	Polish × Min.	1	4799	F_1	Polish × Min.	2	3	7	10	3	2	1	2	..	4	2	..
33	805	5323	F_1	...Do....	1	4799	F_1	...Do....	2	3	17	7	2	1	..	..	2	1	..	..
				Totals (435)							92	147	88	21	22	19	10	13	17	6
				Percentages							21.2	33.8	20.2	4.8	5.0	4.4	2.3	3.0	3.9	1.4
												80.0				20.0				

TABLE 49.—*Distribution of frequency of grades of "openness" in offspring when both parents are heterozygous ($DR \times DR$, F_2 and later generations).*

Serial No.	Pen No.	Mother				Father				Total gr.	Grade of openness in offspring									
		No.	Gen.	Races.	Gr.	No.	Gen.	Races.	Gr.		1	2	3	4	5	6	7	8	9	10
34	763	3799	F_2	Hou.× Wh.L	6	2247	F_2	Hou.× Wh.L	2	8	..	2	2	2	2	..	1	..	..	..
35	765	84	F_1	...Do....	3	1794	F_2	...Do....	5	8	1	6	8	1	4	1	1	2	6	3
36	765	984	F_2	...Do....	3	1794	F_2	...Do....	5	8	8	3	1	2	2	1	0	2	5	..
37	802	4013	F_2	Polish × Min.	4	6652	F_1	Polish × Min.	4	8	6	12	9	6	..	..	..	1	1	1
38	802	3954	F_2	...Do....	3	6652	F_1	...Do....	4	7	4	12	3	2	1	..	2	6	9	1
39	802	4038	F_2	...Do....	3	6652	F_1	...Do....	4	7	3	8	4	3	2	..	1	4	1	1
40	802	4164	F_2	...Do....	3	6652	F_1	...Do....	4	7	6	8	6	2	1	1	..	2	2	2
41	812	84	F_2	Hou.× Wh.L.	3	4118	F_2	Hou.× Wh.L.	4	7	..	1	5	2	2	1	1	1	1	2
42	812	913	F_2	...Do....	3	4118	F_2	...Do....	4	7	10	6	6	1	..	..	..	3	2	5
43	812	4728	F_2	...Do....	3	4118	F_2	...Do....	4	7	8	5	5	1	5	2	2	2	9	2
44	812	5120	F_2	...Do....	3	4118	F_2	...Do....	4	7	1	2	2	..	1	..	..	..	1	2
45	812	5540	F_2	Polish × Min.	3	4118	F_2	...Do....	4	7	2	5	6	1	1	..	..	..	..	..
46	763	2250	F_2	Hou.× Wh.L.	5	2247	F_2	...Do....	2	7	4	10	2	..	..	..	..	1	0	2
47	812	4726	F_2	...Do....	2	4118	F_2	Polish × Min.	4	6	4	6	3	..	2	1	..	2	1	3
48	812	4735	F_2	...Do....	2	4118	F_2	...Do....	4	6	2	1	1	..	..	..	..	2	1	..
49	765	1790	F_2	...Do....	1	1794	F_2	Hou.× Wh.L.	5	6	9	14	9	1	3	0	2	0	3	..
50	802	4012	F_2	Polish × Min.	1	6652	F_1	Polish × Min.	4	5	5	13	11	3	2	..	1	3	1	..
51	825	2198	F_3	...Do....	3	3852	F_2	...Do....	2	5	..	..	1	3	..	..	..	..	..	1
52	728	2271	F_2	Hou.× Wh.L.	3	258	F_1	Hou.× Wh.L.	2	5	4	3	1	7	2	1	3	1	1	2
53	763	2700	F_2	...Do....	3	2247	F_2	...Do....	2	5	1	2	3	2	..	1	..	..	2	..
54	825	350	F_1	Polish × Min.	2	3852	F_2	Polish × Min.	2	4	4	13	6	4	..	..	..	3	1	3
55	825	4708	F_3	...Do....	2	3852	F_2	...Do....	2	4	4	13	7	3	..	1	1	1	2	3
56	825	5019	F_3	...Do....	2	3852	F_2	...Do....	2	4	1	1	..	..	..	..	..	1	2	2
57	825	5035	F_3	...Do....	2	3852	F_2	...Do....	2	4	4	..	3	1	1	..	..	1	1	1
58	825	5672	F_3	...Do....	2	3852	F_2	...Do....	2	4	1	3	2	..	2	..	..	1	2	1
59	728	2248	F_2	Hou.× Wh.L.	2	258	F_1	Hou.× Wh.L.	2	4	3	6	7	2	..	1	0	1	3	..
61	763	377	F_1	...Do....	1	2247	F_2	...Do....	2	3	20	9	14	3	6	0	2	0	2	1
				Totals (663)							115	164	127	53	39	10	8	39	57	41
				Percentages							17.4	24.7	19.2	8.0	5.9	1.5	2.7	5.9	8.6	6.2
												69.3				30.7				

These earlier data were not even roughly quantitative, and it is the quantitative data that first give the key to the true relations. However, sufficient evidence for the change in the conclusion is certainly due. The evidence is found in a careful study of table 55, keeping constantly in mind this fundamental principle that the recessive condition alone in the parents can never give rise to the dominant; for the recessive condition implies entire absence of the dominant factor. But the pure dominant condition will vary in the direction of the recessive condition; such a result implies only a partial failure of the factor to develop completely; and we should not be surprised if occasionally the failure were complete. This implies no "reversal of dominance," but rather an arrested development of the factor.

At the outset, then, we find (table 55) that even pure races with high nostril (Polish, Houdans), when bred together, vary much in the height of nostril (in perfection of dominance) and, in 2 per cent of the offspring,

TABLE 50.—*Distribution of frequency of grades of "openness" in offspring when one parent is heterozygous and the other an original dominant ($DR \times D$, originals).*

Serial No.	Pen No.	Mother.				Father.				Total gr.	Grade of openness in offspring.									
		No.	Gen.	Races.	Gr.	No.	Gen.	Races.	Gr.		1	2	3	4	5	6	7	8	9	10
62	803	529	F₁	Houd.×Min	3	7522	P.	Houd.	9	12	4	2	4	1	2	..	..	2	2	1
63	803	7065	F₁	Houd.×Dk. Brah.	1	7522	P.	...Do...	9	10	6	11	6	4	2	1	2	6	4	1
				Totals (61)							10	13	10	5	4	1	2	8	6	2
				Percentages							16.4	21.3	16.4	8.2	6.5	1.6	3.3	13.1	9.8	3.3
											62.3				37.7					

TABLE 51.—*Distribution of frequency of grades of "openness" in offspring when one parent is heterozygous and the other an extracted dominant ($DR \times DD$, extracted).*

[ABBREVIATIONS: H = Houdan; L = Leghorn; M = Minorca; P = Polish; WL = White Leghorn.]

Serial No.	Pen No.	Mother.				Father.				Total gr.	Grade of openness in offspring									
		No.	Gen.	Races.	Gr.	No.	Gen.	Races.	Gr.		1	2	3	4	5	6	7	8	9	10
64	832	4404	F₂	H×WL	4	5119	F₂	H×WL	10	14	1	1	1	2	..	..	..	8	1	1
65	729	913	F₂	Do.	6	936	F₂	Do.	10	16	5	6	16	2	5	..	3	11	11	10
66	819	57	F₁	P×M	4	1420	F₂	P×M	10	14	3	2	4	..	..	1	..	3	5	1
67	832	505	F₁	H×L	4	5119	F₂	H×WL	10	14	2	2	3	3	2	..	2	2	2	4
68	729	935	F₂	H×WL	4	936	F₂	Do.	10	14	3	5	4	0	3	2	5	12	15	3
69	756	2011	F₂	HPMWL	4	444	F₂	Do.	10	14	..	..	..	1	..	1	..	1	4	3
70	807	185	F₁	P×M	4	3894	F₂	P×M	9	13	4	2	..	..	..	2	1	1	2	1
71	756	1048	F₂	Do.	3	1390	F₂	Do.	10	13	..	..	3	..	..	..	..	..	2	..
72	762	505	..	H×L	3	144	F₂	H×L	10	13	1	..	3	1	..	1	2	2	3	4
73	762	2011	F₂	HPML	4	2621	F₂	HPML	9	13	..	1	..	..	..	..	1	1	3	1
74	813	2271	F₂	H×WL	5	3904	F₂	H×WL	7	12	1	5	5	2	1	3	2	4	..	9
75	820	984	F₁	H×L	3	4731	F₂	P×M	9	12	..	5	4	2	5	1	..	5	5	4
76	728	2272	F₁	Do.	10	258	F₁	H×L	2	12	2	7	9	4	4	3	2	7	7	9
77	756	1043	F₂	P×M	2	1390	F₂	P×M	10	12	5	5	3	2	..	..	..	3	2	2
78	762	505	..	H×L	3	2621	F₂	HPML	9	12	1	..	..	..	..	1	..	..	..	3
79	803	2250	F₂	H×L	3	7522	P.	Houd.	9	12	..	5	2	2	4	..	..	4	9	6
80	803	2254	F₂	Do.	3	7522	P.	Do.	9	12	6	6	4	1	2	1	1	3	6	3
81	769	492	F₁	Do.	2	911	F₂	H×L	9	11	3	6	1	..	2	..	..	..	1	..
82	807	1043	F₂	P×M	2	3894	F₂	P×M	9	11	9	4	2	..	3	3	..	6	6	..
83	769	2254	F₂	H×L	1	911	F₂	H×L	9	10	7	7	2	1	..	..	1	2	4	1
84	813	935	F₂	Do.	3	3904	F₂	Do.	7	10	1	2	..	4	3	..	..	7	8	1
85	813	5113	F₂	Do.	3	3904	F₂	Do.	7	10	4	5	5	..	1	1	1	6	8	5
86	813	5142	F₂	Do.	3	3904	F₂	Do.	7	10	..	2	..	..	..	..	..	1	1	3
87	813	5122	F₂	Do.	2	3904	F₂	Do.	7	9	..	1	2	..	1	..	..	2	2	3
88	813	7320	F₂	Do.	2	3904	F₂	Do.	7	9	..	6	1	..	1	..	..	2	5	2
89	813	377	F₁	Do.	1	3904	F₂	Do.	7	8	10	..	6	1	..	..	1	4	3	..
				Totals (641)							68	86	80	29	38	24	23	95	119	79
				Percentages							10.6	13.4	12.5	4.5	5.9	3.7	3.6	14.8	18.6	12.3
											41.0				59.0					

TABLE 52.—*Distribution of frequency of grades of "openness" in offspring when both parents are extracted dominants (extracted DD×DD).*

[ABBREVIATIONS: H = Houdan; L = Leghorn; M = Minorca; P = Polish; WL = White Leghorn.]

Serial No	Pen No	Mother				Father				Total gr.	Grade of openness in offspring.									
		No.	Gen.	Races.	Gr.	No.	Gen.	Races.	Gr.		1	2	3	4	5	6	7	8	9	10
91	729	2016	F₂	HPLM	10	936	F₂	H×L	10	20	..	..	..	..	..	..	..	4	6	5
92	729	2255	F₂	H×L	10	936	F₂	Do	10	20	3	3	..	..	..	1	2	5	11	10
93	729	2269	F₂	Do	10	936	F₂	Do	10	20	..	1	..	..	..	1	..	3	9	13
94	729	2324	F₂	HPLM	10	936	F₂	Do	10	20	2	3	..	..	1	1	..	5	16	7
95	756	1067	F₂	P×M	10	1390	F₂	P×M	10	20	..	1	3	2	1	..	..	1	1	..
96	756	1113	F₂	Do	10	1390	F₂	Do	10	20	..	..	..	..	..	..	1	4	8	4
97	762	2014	F₂	HPLM	10	444	F₂	H×L	10	20	..	..	..	..	..	..	..	1	4	..
98	819	1113	F₂	P×M	10	1420	F₂	P×M	10	20	..	..	..	..	..	..	..	2	6	2
99	819	4257	F₂	Do	10	1420	F₂	Do	10	20	..	..	..	2	..	..	..	4	4	3
100	832	4732	F₂	H×L	10	5119	F₂	H×L	10	20	..	..	..	..	..	..	..	2	1	..
101	832	6481	F₂	Do	10	5119	F₂	Do	10	20	..	..	..	..	..	..	..	2	5	4
102	756	369	F₂	P×M	9	1390	F₂	P×M	10	19	..	2	..	..	..	..	..	..	1	1
103	762	2618	F₂	HPLM	9	444	F₂	H×L	10	19	..	..	..	..	..	..	..	1	1	..
104	762	3776	F₂	H×L	9	444	F₂	Do	10	19	..	..	..	..	..	..	..	1	1	..
105	832	5803	F₂	Do	9	5119	F₂	Do	10	19	..	..	1	1	..	..	1	6	9	6
106	807	1067	F₂	P×M	10	3894	F₂	P×M	9	19	..	1	1	1	2	1	2	1	4	2
107	762	2333	F₂	HPLM	8	444	F₂	H×L	10	18	..	..	..	..	..	..	1	2	5	4
108	762	2618	F₂	Do	9	2621	F₂	HPLM	9	18	..	..	..	..	..	..	1	2	2	..
109	762	3776	F₂	H×L	9	2621	F₂	Do	9	18	..	..	..	..	..	1	..	2	4	4
110	819	5674	F₂	P×M	8	1420	F₂	P×M	10	18	1	..	1	..	..	..	2	1	3	2
111	820	2016	F₂	HPLM	9	4731	F₂	Do	9	18	..	..	..	..	1	..	1	1	4	..
112	820	2255	F₂	H×L	9	4731	F₂	Do	9	18	..	..	..	..	..	..	1	2	6	5
113	820	6479	F₂	Do	9	4731	F₂	Do	9	18	..	..	1	..	2	1	2	9	12	4
114	832	2618	F₂	HPLM	8	5119	F₂	H×L	10	18	1	1	..	4	..	..	..	..	12	3
115	832	3776	F₂	H×L	8	5119	F₂	Do	10	18	..	3	3	..	2	..	..	..	..	..
116	834	2324	F₂	HPML	9	5090	F₂	Do	9	18	..	1	..	..	..	1	..	10	10	3
117	762	2333	F₂	HPLM	8	2621	F₂	HPLM	9	17	..	..	..	..	1	..	1	..	..	1
118	807	5075	F₂	P×M	7	3894	F₂	P×M	9	16	..	1	..	..	2	1	..	5	7	7
119	820	5143	F₂	H×L	7	4731	F₂	Do	9	16	1	..	2	5	..	1	3	10	10	12
120	813	2272	F₂	Do	9	3904	F₂	H×L	7	16	1	1	1	..	1	..	2	5	7	7
		Totals (472)									9	19	18	13	14	8	22	93	169	105
		Percentages									1.9	4.0	3.8	2.8	3.0	1.7	4.7	19.8	36.0	22.3

TABLE 53.—*Distribution of frequency of grades of "openness" in offspring when both parents are heterozygous (RR×DR).*

Serial No.	Pen No.	Mother				Father				Grade of openness in offspring.									
		No.	Gen.	Races.	Grade.	No.	Gen.	Races.	Grade.	1	2	3	4	5	6	7	8	9	10
121	728	174	F₁	Houd.×Legh.	1	1258	P.	Brab.×Tosa.	2	2	7	2	1	..	1	1	..	..	..
122	728	912	F₂	Do	2	258	F₁	Houd.×Legh.	2	7	3	3	2	1	..	..	..	..	..
123	763	3799	F₂	Min.×Houd.	6	2247	F₂	Do	2	..	2	2	2	2	..	1	..	2	..
124	802	509	F₂	Polish×Min.	1	6652	F₁	Polish×Min.	4	6	6	1	..	1	..	..	..	..	..
125	802	3846	F₂	Do	2	6652	F₁	Do	4	1	6	3	1	1	..	..	..	..	..
126	802	5025	F₂	Do	2	6652	F₁	Do	4	8	10	4	3	2	..	..	..	..	..
127	802	5087	F₂	Do	2	6652	F₁	Do	4	7	9	12	2	..	1	..	..	..	1
		Totals (217)								31	43	27	11	8	2	2	0	2	1
		Percentages								24.4	33.9	21.3	8.7	6.3	1.6	1.6	0	1.6	0.8

TABLE 54.—*Distribution of frequency of grades of "openness" in offspring when both parents are extracted recessives (extracted RR×RR).*

Serial No.	Pen No.	Mother				Father				Total grade.	Offspring.	
		No.	Gen.	Races.	Grade.	No.	Gen.	Races.	Grade.		Grade 1	Grade 2
128	728	*912	F₂	Houd.×Legh.	2	1298	F₂	Houd.×Legh.	1	3	3	3
129	827	298	F₂	Pol.×Min.	2	3852	F₂	Do	2	4	5	5

* *Cf.* Serial No. 12a.

even show the typical narrow nostril (fig. B, a). On the other hand, in the narrow-nostriled races I have never obtained any such variation. The most deviation that I have seen from grade 1 is found in my strain of Dark Brahma bantams that frequently give grade 2. The variability of the high nostril, the stability of the low nostril, is *prima facie* evidence that the former is due to the presence of a particular factor and the latter to its absence.

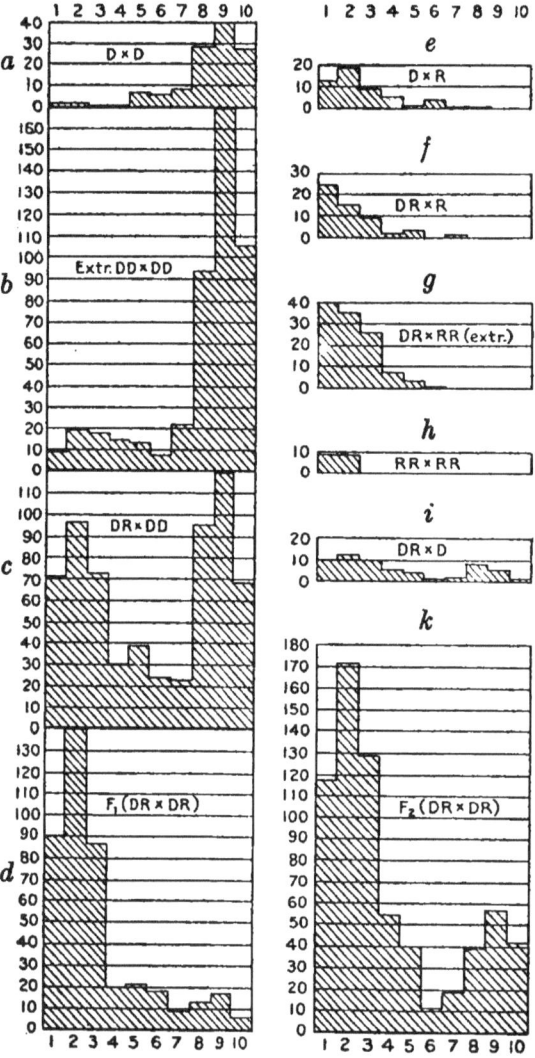

Fig. B.—Polygons of frequency of grades of "openness" of nostril in offspring of various parents. a, Both parents pure bred dominants; b, both parents extracted dominants; c, one parent heterozygous, the other a dominant; d, both parents heterozygous; e, dominant by recessive; f, heterozygous by recessive; g, heterozygous by extracted recessive; h, extracted recessives; i, heterozygous by dominant; k, both parents second generation hybrids.

Next, the heterozygotes of F_1 (table 46), may be appealed to; but they will give no critical answer. For expectation, dominance being imperfect, is that the hybrids will be intermediate, and the result will be the same whichever extreme grade is taken as dominant. The empirical mode in the distribution of the offspring is at grade 2. This implies much greater imperfection of dominance on the hypothesis that grade 10 is dominant than on the hypothesis that grade 1 is dominant; but this very fact supports the former hypothesis, since imperfection of dominance is obviously a feature of the character with which we are dealing.

The critical test is afforded by the F_2 generation (tables 48 and 49). By hypothesis, 25 per cent of the offspring are expected to be pure ("extracted") recessives, and the same number pure dominants; and also, by hypothesis, the recessives are massed at or near one grade while the dominants are variable. Now, as a matter of fact, the upper 25 per cent range over 5 to 7 grades, while the lower 25 per cent are nearly massed in grade 1

(21 per cent are so massed in one table, 17 per cent in the other). Therefore, in accordance with hypothesis we must regard the lower grade—narrow slit—as recessive. Similarly, heterozygous × low nostril (table 47) should give, on our hypothesis, 50 per cent low nostril. If that is recessive we should expect a massing of this 50 in the first two grades; if dominant a greater scattering. The former alternative is realized. Again, in the heterozygous × high nostril hybrid (table 50) the upper 50 per cent will be massed or scattered according as high nostril is recessive or dominant. Allowing for the 50 per cent heterozygotes in the progeny, the 50 per cent of high nostrils are scattered through at least 8 grades of the possible 10. High nostril is dominant. Finally, extracted high nostrils bred together produce offspring (table 52) with a great range of variability (through all grades), while extracted low nostrils (unfortunately all too few) give progeny with grades 1 and 2 (table 53; fig. B, h). Accepting, then, the general principle of the greater variability of the dominant character, we have demonstrated conclusively that high nostril, or rather the factor that determines high nostril, is dominant.

Comparing tables 45 to 54, we see that recessive parents are characterized by a low grade of nostril and they, of course, tend to produce offspring with a low grade. Similarly, dominants have a high grade and tend to produce offspring of the same sort, while heterozygous parents are of intermediate grade and their children have nostril grades that are, on the average, intermediate. Without regarding the gametic constitution, we might conclude, with Castle, that offspring inherit the grade of their parents, and consequently it would be possible to increase the grade, perhaps indefinitely, by breeding from parents with the highest grade. Considering the gametic constitution of the parents, it is obvious that such a conclusion is premature. To get an answer to the question it is necessary to find if there is, inside of any one table, among parents of the same gametic constitution, any such relation between parental and filial grades. This can be determined by calculating the correlation between the grades of parents and progeny. Such calculation I have made for table 48 with the result: index of correlation, $r = 0.018 \pm 0.032$, which is to be interpreted as indicating that no correlation exists; and in so far the hypothesis of Castle proves not to apply in the cases of booting and doubt is thrown on the significance of his conclusion.

Finally, if we throw together the frequency distributions of all tables into one table (table 55; compare fig. B) we shall find the totals instructive. Table 55 shows that, when all results are thrown together, including hybrids of all sorts, grade 2 and grade 9 are the most frequent and grade 6 is the least frequent, the frequency gradually rising towards the extremes of the series. The same result appears in the individual series that range from grade 1 to grade 10. What is the meaning of this result? It seems to me to bear but one interpretation, namely, that there are only two centers of stability—about grades 1 and 9—and true blending of these grades, giving an intermediate condition, does not occur. Otherwise, in consequence of the repeated hybridization, the intermediate grades must be the commonest instead of the rarest. There is alternative inheritance of the nostril height.

TABLE 55.—*Summary of tables 45 to 54.*

Table No.	Nature of mating (parental nostril).	Nature of mating.	Grade of openness in offspring.										
			1	2	3	4	5	6	7	8	9	10	Total
ABSOLUTE FREQUENCIES.													
45	High × high	D×D	2	2	1	1	6	5	8	28	39	27	119
46	High × low	D×R	13	19	9	7	2	4	1	1	...	...	56
47	Heterozygous × low	DR×R	23	14	9	2	4	...	1	...	...	...	53
48	Heterozygous × heterozygous	DR×DR	90	140	86	20	21	18	9	13	17	6	420
49	....Do	F₂(DR×DR)	117	171	129	54	40	11	19	39	57	41	678
50	Heterozygous × high	DR×D	10	13	10	5	4	1	2	8	6	2	61
51	....Do	DR×DD	71	96	73	30	39	24	23	95	119	68	638
52	Extra high × high	DD×DD	9	19	18	15	14	8	22	93	169	105	472
53	Heterozygous × extracted low	DR×RR	40	35	26	7	3	1	...	...	...	...	112
54	Extra low × low	RR×RR	8	8	...	...	...	...	...	...	...	...	16
	Totals		378	512	361	141	133	72	85	277	407	249	...
PERCENTAGES.													
45	High × high	D×D	1.7	1.7	0.8	0.8	5.0	4.2	6.7	23.5	32.8	22.7	....
46	High × low	D×R	23.2	34.0	16.1	12.5	3.6	7.1	1.8	1.8	....	....	....
47	Heterozygous × low	DR×R	43.4	26.4	35.9	3.8	7.6	....	1.9	....	....	....	....
48	Heterozygous × heterozygous	DR×DR	21.5	33.3	20.5	4.8	5.0	4.3	2.1	3.1	4.1	1.2	....
49	....Do	F₂(DR×DR)	17.3	25.2	19.0	8.0	5.9	1.6	2.8	5.8	8.4	6.1	....
50	Heterozygous × high	DR×D	16.4	21.3	16.4	8.2	6.6	1.6	3.3	13.1	9.8	3.3	....
51	....Do	DR×DD	11.1	15.1	11.4	4.7	6.1	3.8	3.6	14.9	18.7	10.7	....
52	Extracted high × high	DD×DD	1.9	4.0	3.8	3.2	3.0	1.7	4.7	19.7	35.8	22.2	....
53	Heterozygous × extracted low	DR×RR	35.8	31.3	23.2	6.3	2.7	0.9	....	....	....	....	....
54	Extracted low × low	RR×RR	50.0	50.0	....	....	....	....	....	....	....	....	....

CHAPTER VIII.

CREST.

In my report of 1906 I called attention to the nature of inheritance of the crest in the first and second generations. The result seemed simple enough on the assumption of imperfect dominance. However, in later generations difficulties appeared, one of which was referred to in a lecture given before the Washington Academy of Sciences in 1907. I stated (1907, p. 182), that "when a crested bird is crossed with a plain-headed one, and the crested hybrids are then crossed *inter se*, the extracted recessives of the second hybrid generation are plain-headed, to be sure, but they show a disturbance of certain feathers." This was an illustration of the statement that recessives which are supposed to come from two pure recessive gametes show in their soma traces of the dominant type. Dr. W. J. Spillman, who was present, made the suggestion that the crest is composed of two characters, T and t, instead of a simple element, and that when t alone is present the result will be the roughened *short* feathers on top of the head.

Further studies demonstrate the validity of this suggestion. There are in the crest two and probably three or more factors. There is a factor that determines length of the feathers and a factor that determines their erectness. There is probably also an extension factor that controls the area that the crest occupies on the head. Thus flatness of position dominates over its absence (or erectness). This is seen even in the first generation. Figs. 5, 6, 8, and 17 of my report of 1906 show this very plainly. They also show that continued growth of feather is dominant over interrupted growth. Thus in the second hybrid generation I got birds with short and erect feathers and one of these is shown in fig. 11 of the 1906 report. That shortness is recessive is proved by various matings of extracted short × short crest. Of 29 offspring none have a higher grade than 1, grade 10 being of full length. On the other hand, two parents with long feathers in the crest (grades 6 to 8) give 5 offspring of grade 1, 12 of grades 5 to 10, thus approaching the 1 : 3 ratio expected from two DR parents. That erectness is recessive is proved by various matings of extracted erect × erect crest. Of 25 offspring none has a lower grade than 4 (1 case) or 5 (1 case). On the other hand, two parents with extracted non-erect feathers give in 46 offspring 13 with feathers whose grade of erectness is 6 or higher and 33 with a grade of 5 or below—of these half of grade 0—close to the expected 1 : 3. The evidence is conclusive that there are two factors in crest that behave in Mendelian fashion—a factor determining the prolonged growth of the feather and a factor causing the feathers to lie repent.

The area of the head occupied by the crest is also variable. This was estimated in tenths for each of the parents and offspring. Two principal centers of variation appeared, at 3 and at 8, or roughly one-third and two-thirds the full area. The results, being based on estimates, are not wholly satisfactory, but so far as they go they indicate that when both parents have a crest that belongs to the lower center of variation their offspring belong chiefly if not exclusively to that center; but when they both belong to the upper center of variation a minority of the offspring belong to the lower center. Provisionally it may be concluded that extensive crest is dominant over the restricted crest or that there is an "extension factor."

CHAPTER IX.

COMB-LOP.

In races having a large single comb this is usually erect in the male, but in the female lops over to the right or left side of the head. This lop is determined before hatching; indeed, in the male it may be ascertainable only in the embryo or in the recently hatched chick. The position of the comb is permanent throughout the life of the pullet and hen and, if pressed to the opposite side, it quickly returns to its original position. At one time I entertained the hypothesis that its position was determined by the pressure of the foot against the head while the chick was still within the shell; but after finding the comb lying both to the right and to the left when in contact with the foot I abandoned this hypothesis as untenable. It seemed possible that this position is hereditary, and so data were collected to test this hypothesis. It is not always easy to decide definitely, even for the female, as to the direction of the lop; for the anterior part of the comb may lop to the right, the posterior part to the left, or *vice versa*. In that case one selects the larger or better defined lopping portion to designate as *the* lop. This is usually the posterior portion of the comb. However, such doubtful cases may be omitted from consideration here, as there are plenty of examples of well-defined lop on both sides of the head.

TABLE 56.

Both parents with right lop.					Mother left lop, father right.				
Pen No.	No. of mother.	No. of father.	Offspring.		Pen No.	No. of mother.	No. of father.	Offspring.	
			Right.	Left.				Right.	Left.
817	6188	3900	7	8	831	1980	4213	9	17
817	6406	3900	12	17	904	3901	7840	4	3
831	1011	4213	7	16	904	7645	7840	6	3
831	3040	4213	13	10				19	23
831	4219	4213	4	21	Mother right lop, father left.				
831	6602	4213	6	15	903	3946	8463	2	0
833	1310	4222	4	7	903	4079	8463	7	2
833	4361	4222	6	4	903	4082	8463	11	6
833	7519	4222	2	4				20	8
904	4714	7870	6	7					
			67	109					

Both parents with left lop.					Summary.				
					Parents.		Offspring.		
						Total.	Right.	Left.	
841	3867	3890	3	9			P. ct.	P. ct.	
841	4663	3890	9	7	Both with right lop	176	38	62	
903	9824	8463	6	5	Both with left lop	39	46	54	
			18	21	Mother left lop, father right	42	45	55	
					Mother right lop, father left	28	71	29	

From table 56 it appears, summing all cases, that there are more left-lopping offspring than right-lopping as 161 to 124 or as 56.5 per cent to 43.5 per cent and that this excess holds whether both parents are right-lopping, or both left-lopping, or the mother left and the father right. Only in the case when the mother is right-lopping is there a majority of offspring of the same sort, but here the numbers are too inconsiderable to carry much weight. Although there is not clear evidence of any sort of inheritance, it is probable that the position of the lop is not determined by a single factor, but by a complex of factors.

The conclusion that right and left conditions are not simple, alternative qualities accords with the results obtained by others. Thus Larrabee (1906) finds that the dimorphism of the optic chiasma of fishes (in some cases the right optic nerve being dorsal and in others the left) is not at all inherited, but in each generation the result is strictly due to chance. This is, perhaps, the same as my conclusion that the hereditary factors are complex. Lutz (1908) finds that in the mode of clasping the hands interdigitally the right thumb is uppermost in 73 per cent of the offspring when both parents clasp with right thumb uppermost, but in only 42 per cent of the offspring when both parents clasp with left thumb uppermost. The mode of clasping is inherited, but not in simple Mendelian fashion.

CHAPTER X.

PLUMAGE COLOR.

A. THE GAMETIC COMPOSITION OF THE VARIOUS RACES.

Plumage color, like hair color, varies greatly among domesticated animals. This diversity is, no doubt, in part due to the striking nature of color variations, but chiefly to the fact that the requisite variations are afforded in abundance. The principal color varieties, in poultry as in other domesticated animals, are melanism, xanthism, and albinism. In addition, poultry show the dominant white, or "gray" white, first recognized in poultry by Bateson and Saunders (1902), which is also found in many mammals, as, for instance, in goats, sheep, and cattle. Besides these uniform colors, we find numerous special feather-patterns, such as lacing (or edging of the feather), barring, penciling, and spangling. Also, there are special patterns in the plumage as a whole, such as wing-bar, hackle, saddle, breast, and top of head (crest). Now, all of these color characters are inherited each in its own definite fashion.

In studying the color varieties of poultry we must first of all, as in flower color (Correns, 1902), mice (Cuénot, 1903), guinea-pigs and rabbits (Castle), various plants and animals (Bateson and his pupils), recognize the existence of certain "factors." In poultry the factors that I have determined are as follows:

C, the color factor, absence of which results in albinism.
J, the Jungle-fowl pattern and coloration.
N (nigrum), the supermelanic factor.
X, the superxanthic or "buff" factor.
W, the graying (white) factor.

We have now to consider how these factors are combined in birds of the different races.

1. WHITE.

Albinos.—These seem to be of two different origins:[*] White Cochins and white Silkies. The white Silkies that I have studied have the gametic formula $cJnwx$; *i. e.*, they have the Jungle-fowl marking, but lack the "color enzyme," supermelanic coat, the graying factor, and the xanthic factor.

"*Grays.*"—White Leghorns and their derivatives belong to this class. Its gametic formula is: $CJNWx$. This indicates that the race contains the

---

[*] Bateson and Punnett (1908, p. 28) recognize three "kinds" of recessive whites—that of the Silkie, that of the Rose-comb bantams, and that of "white birds that have arisen in the course of our experiments." White Cochins have perhaps been one of the ancestors of Rose-comb bantams; Bateson's new white lay recessive in the White Dorking and when mated to the White Silkie throws Game-colored offspring.

color enzyme, as well as the Jungle pattern and the supermelanic coat. But all of these are rendered invisible by the graying factor W. The superxanthic factor is missing.

2. BLACK.

The uniform black birds that I have studied are of several sorts. The Black Minorca and White-faced Black Spanish have the gametic formula $CJNwx$. Owing to the absence of the graying factor and the presence of the color factor these appear as pigmented birds, but the supermelanic coat, N, obscures the Jungle coloration, so that the bird appears entirely black. Nevertheless the black is not of uniform quality, but just those parts of the feathers of the wing, back, hackle, saddle, and breast that are red in the Jungle fowl are of an iridescent black, while the portion that is not red in the Jungle is of a dead black.

The Black Cochin has the gametic formula $CINwx$. This differs from the formula of the Minorca only in this respect: the Jungle pattern is present, but not the pigmentation that is usually associated with it.

The Black Game ("Black Devil") that I used in a few experiments seemed to have the same gametic formula as the Minorca, only the supermelanic coat was less dense.

3. BUFF.

For this color I used Buff Cochins, the original buff race. The gametic formula of this race proves to be $CjnwX$—the Jungle-fowl pattern being absent.

B. EVIDENCE.

The evidence for the gametic interpretations of the self-colored fowl is derived from hybridizations. It will now be presented in detail.

1. SILKIE × MINORCA (OR SPANISH).
(Plates 3 to 6.)

By hypothesis this cross is between $cJnwx$ and $CJNwx$. The first generation should give the zygotic formula $CcJ_2Nnw_2x_2$, or, more simply, CcJ_2Nn. This formula resembles closely that for the Minorca; but it differs in this important respect, that the coloring factor and the supermelanic factor are both heterozygous, and hence diluted.

Actually I found, as Darwin (1876) did, that the chicks of this first hybrid generation were all wholly black. In this respect they differed markedly from the chicks of the Silkie, which are pure white, and also from the chicks of the Minorca, which are prevailingly black, but have white belly and outer primaries. The white in the young chicks of Minorcas is extremely variable in amount, but never wholly absent; in time, as the bird grows older, it is replaced by black, so that the adult male and female Minorcas have a wholly black plumage. The reason for the precocious development of black pigment over the belly and primaries of the hybrid chicks is probably the presence of an extension factor (cf. Castle, 1909) derived from the Silkie. Certain it is that the ordinary Jungle pattern develops pigment on the belly and on the wings, as well as on other parts of the plumage. The hybrid chicks may be said to have the extended pigmentation dominant over interrupted pigmentation. In the adult hybrids a difference appears between the coloration of the male and female, even as Darwin pointed out. For the latter retains its uniform blackness, while the former gains red on the wing-bar, and saddle and hackle lacing (plate 4). Now, since all the factors present in the Minorca, and none others, are present in the hybrids, why should the male hybrids show red, and why should the males show red and not the females? The answer to the first question is, I think, clear. While the Jungle pattern of black and red is completely obscured by the undiluted N factor of the Minorca, it is only incompletely covered by the diluted, heterozygous N factor of the hybrid. Hence the red appears in greatly reduced amount, as compared with the Jungle-fowl. In the female Jungle-fowl there is little red and consequently none shows in the female hybrid. Thus the difference in the sexes of the hybrids corresponds to the sexual dimorphism of the Jungle-fowl; but the hybrids are, as indicated, very unlike the Jungle-fowl in coloration (cf. plates 1 and 2).

Since segregation takes place in the gametes of these heterozygotes, 4 kinds of gametes are possible, namely, CJN, CJn, cJN, cJn. On mating heterozygotes together, zygotes of 16 types will be formed, as in table 57.

TABLE 57.—*Zygotes in F_2 of Silkie × Minorca hybrids and their corresponding somatic colors.*

$C_2J_2N_2$	N	C_2J_2Nn	N	CcJ_2N_2	N	CcJ_2Nn	N
C_2J_2Nn	N	C_2J_2nn	G	CcJ_2Nn	N	CcJ_2nn	G
C_2JjN_2	N	C_2JjNn	N	cjJ_2N_2	W	cjJ_2Nn	W
C_2JjNn	N	C_2Jjnn	G	cjJ_2Nn	W	$cjJjnn$	W

In the foregoing table there is given after each combination a letter: N standing for black, the appearance of the soma; G standing for Game-colored, and W standing for white. No distinction is made between pure blacks and those that, while black as chicks, subsequently show some red in the male. Such a distinction was impracticable because most of the color

TABLE 58.

Pen No.	Black.		White.		Game.	
	Observed.	Expected.	Observed.	Expected.	Observed.	Expected.
709.....	119	116	55	51	31	38
804.....	91	89	40	39	26	29
Total....	210	205	95	90	57	67

determinations are made on the young chicks. It appears that in 16 progeny expectation is 9 black, 4 white, and 3 Game-colored. Actually 362 offspring were obtained, with the results shown in table 58. Nothing is more striking than to see the hens of this F_2 generation with evidences of the female Game pattern (plate 6).

Comparing observed results in the distribution of colors in the F_2 generation with expectation, it is seen that the proportions are close, and this closeness of observation with expectation is evidence for the correctness of the hypothesis.

The hypothesis may be further tested in later generations by breeding together the different sorts of individuals obtained in F_2. In pursuance of such a test I mated various pure black hens with pure black cocks and those of F_1, and, as was to have been expected, obtained families of different sorts, simply because even pure blacks have differing gametic constitutions. Thus in pen 824 I mated an extracted black cock with 3 black hens. All were apparently of the zygotic constitution C_rJ_sNn, forming gametes CJN and CJn. Mated together these should give the three black combinations $C_rJ_sN_s$, C_rJ_sNn, C_rJ_snN, to one Game, $C_rJ_sn_s$. Actually there were obtained 64 black and 23 Game, 66 to 22 being expectation. In another pen (pen 804) an F_1 cock was mated to various black F_2 hens. The families fall into 2 classes. The cock, of course, produced gametes CJN, CJn, cJN, cJn. With four females like him (Nos. 3902, 3908, 5431, 6043) I got: black 40, white 13, Game 14; expected, black 38, white 17, Game 13. Three females (Nos. 4715, 4716, 5099) evidently produced gametes CJN, CJn. Expectation is that blacks and Games shall be produced in the proportions of 3 to 1. Actually 30 : 14 were obtained where 33 : 11 was expected. All of these results accord closely with the hypothesis.

The whites obtained in F_2 are of 3 types, but in all alike the color factor is missing. Hence it can not reappear in the offspring, and, consequently, no colored offspring are to be expected. But, first, it must be stated that the extracted whites of the F_2 generation are not always of a pure white. Indeed, the parent Silkies are in some cases not perfectly white, but show traces of "smoke." There are different degrees of albinism; the coloring enzyme may be absent to small traces. This variability in degree of albinism is familiar to all students of albinism in man. My breeding of extracted whites was done in pen 817 and consisted of a pure white cock (No. 3900) and 2 hens. Of these 1 (No. 6046) was pure white and produced in a total of 15 only white offspring, but among those that were described as unhatched I have recorded traces of pigment in 24 per cent of the cases. The second hen (No. 3899) had black flecks in the white plumage. She had 20 offspring, of which 2 (unhatched) are recorded as having N down, 2 as "blue," and 3 others show traces of black pigment. Thus, 7 birds in 20, or 35 per cent of all, show more or less black, even as the albinic mother does. On the whole, however, omitting from present consideration the phenomenon of incomplete albinism, we may say that 2 pure albino parents produce only albinic offspring, while imperfectly albinic parents produce some imperfectly albinic offspring.

2. SILKIE × WHITE LEGHORN.

By hypothesis this cross is between $cJnwx$ and $CJNWx$. The first generation should give the zygotic formula CcJ_2NnWwx_2, or, more simply, CcJ_2NnWw. This formula resembles closely that of the White Leghorn, except that the coloring and graying factors and that for supermelanism are all heterozygous and hence diluted; only the Jungle coloration remains unchanged. Actually, the first generation yielded a lot of white birds like the Leghorn, but with this difference, that, as the males became mature, they gained red on the wing-bar and to a slight extent on the lacing of the saddle. The females gained a faint blush of red on the breast. Thus red appeared, in small amount, in just those places in the respective sexes which are red in the Jungle-fowl. The explanation of its appearance that I have to suggest is that, both on account of the diluting of the supermelanic coat and of the graying factor, the red of the undiluted underlying Jungle coloration is revealed.

Since the hybrids are heterozygous in respect to 3 pairs of characters, when segregation occurs each parent produces 8 kinds of gametes, as follows: $CJNW, CJNw, CJnW, CJnw, cJNW, cJNw, cJnW, cJnw$. When both parents produce these 8 kinds of gametes we may expect, in 64 offspring, the proportions of the several types shown in table 59.

TABLE 59.—*Probable frequency in 64 progeny.*

Zygotic formula.	White.	White + red.	Game.	Black.	Zygotic formula.	White.	White + red.	Game.	Black.
$C_2J_2N_2W_2$	1	..	..	..	CcJ_2N_2Ww	4	..	..	..
$C_2J_2N_2Ww$	2	..	..	..	$CcJ_2N_2w_2$	..	..	..	2
$C_2J_2N_2w_2$	..	..	..	1	CcJ_2NnW_2	4	..	..	..
$C_2J_2NnW_2$	2	..	..	..	CcJ_2NnWw	..	8	..	..
C_2J_2NnWw	..	4	..	..	CcJ_2Nnw_2	..	..	4	..
$C_2J_2Nnw_2$	..	..	2	..	$CcJ_2n_2W_2$	2	..	..	..
$C_2J_2n_2W_2$	1	..	..	..	CcJ_2n_2Ww	..	4	..	..
$C_2J_2n_2Ww$	..	2	..	..	$CcJ_2n_2w_2$	..	..	2	..
$C_2J_2n_2w_2$	..	..	1	..	c_2J_2—.	16	..	..	..
$CcJ_2N_2W_2$	2	..	..	..	Total (64)..	34	18	9	3

While, if the progeny were all to survive to maturity, we might expect to get the proportions of white and of white-and-red progeny called for, yet, since the red color appears in most cases at an age *after* the chicks are described, it will be necessary in comparing experience with calculation to combine the first two classes as whites. We then find the proportions given in table 60.

TABLE 60.

Color.	In 64, calculated.	In the actual 85 individuals.	
		Calculated.	Observed.
White	52	69	68
Game	9	12	16
Black	3	4	1

The proportion of whites agrees closely with expectation. If this is not the case with the other two classes, the discrepancy must be attributed in part to insufficient observations and in part to the difficulties of precise classification in the early stages. The result is so close, however, as to lend strong support to our hypothesis as to the gametic constitution of the parents. Nothing is more striking, and to the unprejudiced mind more convincing, than the appearance of typically Game-colored birds in the grandchildren of wholly white parents.

3. SILKIE × BUFF COCHIN.
(Plates 7, 8.)

By hypothesis this cross is between $cJnwx$ and $CjnwX$. The first generation should give the zygotic formula $CcJjn_2w_2Xx$, or, more simply, $CcJjXx$. The formula differs much from that of either parent, and the progeny themselves are no less remarkable. They have a washed-out buff color (since they are heterozygous in both C and X), and the Jungle pattern shows itself in the black tail and slightly redder buff of the wing-bar and hackles in the male. Since the hybrids are heterozygous in respect to 3 pairs of characters, when segregation occurs each parent produces 8 kinds of gametes, as follows: CJX, CJx, CjX, Cjx, cJX, cJx, cjX, cjx. In F_2 the types listed in table 61 may be expected in 64 offspring.

TABLE 61.—*Distribution of colors, theoretic classes.*—*Probable frequency in 64 progeny.*

Zygotic formula.	White.	Buff.	Buff + black.	Game.	Zygotic formula.	White.	Buff.	Buff + black.	Game.
$C_2J_2X_2$			1		CcJ_2Xx			4	
C_2J_2Xx			2		CcJ_2xx_2				2
$C_2J_2xx_2$				1	$CcJjX_2$			4	
C_2JjX_2			2		$CcJjXx$			8	
C_2JjXx			4		$CcJjxx_2$				4
C_2Jjxx_2				2	$Ccjxx_2$		2		
C_2jjX_2		1			Ccj_2Xx		4		
C_2jjXx		2			$Ccjxx_2$	2			
C_2jjxx_2	1				cc—	16			
CcJ_2X_2			2						
					Total. .	19	9	27	9

The classification here employed can not be used in detail in comparing observed results with expectation, for the distinction between buff and buff-and-black appears only in chicks that have acquired the permanent plumage. Consequently it will be found necessary to combine these two classes into one and then make the comparison—as is done in table 62.

TABLE 62.—*Distribution of colors, combined classes.*

Color.	In 64, calculated.	In the actual 58 individuals.	
		Calculated.	Observed.
Buff (and black)	36	33	34
White	19	17	17
Game	9	8	7
Total	64	58	58

The correspondence is certainly close. The hypothesis of factors thus receives additional support and the variability of the offspring in the second hybrid generation is sufficiently explained.

4. WHITE LEGHORN × BLACK MINORCA.

As we have already seen, the gametic formula of the White Leghorn is $CJNWx$ and that of the Minorca is $CJNwx$, so that the F_1 generation has the zygotic formula $C_2J_2N_2Wwx_2$ or, more simply, $C_2J_2N_2Ww$. These heterozygotes are white because of the graying factor, but, as this factor is diluted, some black shows, particularly in the females. In F_2, on account of there being only 1 heterozygous factor, only 3 kinds of zygotes are formed, $C_2J_2N_2W_2$, $C_2J_2N_2Ww$, and $C_2J_2N_2w_2$, in the proportions 1 : 2 : 1. Since not only offspring homozygous in W, but also all male heterozygotes, are white and many female heterozygotes are late in revealing any pigment, it is necessary to consider together individuals homozygous and heterozygous in W. Consequently we may expect 75 per cent of the offspring to show white or white-black-speckled plumage, and 25 per cent black or black and white like the young Minorca. Actually, in 154 offspring (pen 633) I obtained 116 white + white-black + blue, and 38 black with more or less white and including 4 barred, of which more later. Expectation is 115.5 and 38.5, respectively.

In another experiment I crossed the F_1 hybrids on a pure White Leghorn and got 41 offspring, all white except 1 that showed some black specks. All results thus accord with hypothesis.

5. WHITE LEGHORN × BUFF COCHIN.
(Plate 9.)

These two races afford the gametic formulæ $CJNWx$ and $CjnwX$, respectively. The F_1 hybrids consequently have the zygotic formula $C_2JjNnWwXx$. Such hybrids are heterozygous in all factors except C. Such complex heterozygotism, combined with the well-known sex differences in color of heterozygotes, leads to a very great diversity of the offspring. As a matter of fact I found, as Hurst did, that the young were sometimes quite white, sometimes white and buff, and sometimes showed also a little black. Since there are 4 heterozygous characters, there are 256 possible combinations of them, which reduce to 81 different kinds of combinations. Owing to the ambiguous nature of the soma in many of the heterozygotes and to the relatively small number of offspring, it is useless to compare theoretical and observed distributions of plumage colors in the somas. Suffice it to say that white, buff, black, and Game-colored chicks all appeared in the F_2 generation, as well as some with a mixture of colors, as called for by the hypothesis. White, due to the powerful graying factor, was the commonest color, buff and black were about equally common, and each about one-third as abundant as white, while Games, due to the hypostatic J factor, were about one-third as common as buff. All this, again, is explicable upon our hypothesis and upon none other so far proposed.

In mating the F_2 generation with each other or with the White Leghorn the result must vary with the gametic output of the hybrid, which is obviously very different in different cases. A hen, of a light buff color spangled with white spots and having a black tail, presumably formed gametes $CJnWX$, $CJnwX$, $CJNWX$, $CJNwX$. Mated with the White Leghorn, $CJNWx$, she produced 8 pure whites, 4 whites with some black and red, 2 buff and white, and 3 black with trace of white. Expectation in 16 offspring would be about 4 pure whites, 4 white mixed with pigment, 4 buffs with white (and black?), and 4 blacks mixed with other colors. This is merely an illustration of the way the confused combinations of colors become intelligible, and even necessary on the factor hypothesis.

6. BLACK COCHIN × BUFF COCHIN.
(Plate 10.)

The factors involved in this cross seem to be $CINx$ for the Black Cochin (in which I stands for the Jungle pattern without any associated color factor) and $CjnX$ for the Buff Cochin, as before. The F_1 generation has the zygotic composition $C_iIjNnXx$, and the females are all black, except for a variable amount of red on the hackle, and the males are black and red, like Games. The F_2 generation is remarkable. Since 3 factors are heterozygous, there are 64 possible combinations and 27 differing ones. In table 63 is given a list of these different combinations and of the probable associated somatic colors. The prefixed number indicates the frequency of each combination.

TABLE 63.

1....$C_2I_2N_2X_2$.....Black.	2....$C_2IiN_2X_2$.....Black.	1....$C_{ii}I_2N_2X_2$.....Black.
2....$C_2I_2N_2Xx$.....Black.	4....C_2IiN_2Xx.....Black.	2....$C_{ii}I_2N_2Xx$.....Black.
1....$C_2I_2N_2x_2$.....Black.	2....$C_2IiN_2x_2$.....Black.	1....$C_{ii}I_2N_2x_2$.....Black.
2....$C_2I_2NnX_2$.....Black and red.	4....C_2IiNnX_2.....Black and red.	2....$C_{ii}I_2NnX_2$.....Black and red.
4....C_2I_2NnXx.....Black.	8....$C_2IiNnXx$.....Black.	4....$C_{ii}I_2NnXx$.....Black.
2....$C_2I_2Nnx_2$.....Black.	4....C_2IiNnx_2.....Black.	2....$C_{ii}I_2Nnx_2$.....Black.
1....$C_2In_2X_2$.....Buff.	2....$C_2Iin_2X_2$.....Buff.	1....$C_{ii}In_2X_2$.....Buff.
2....C_2In_2Xx.....Buff.	4....C_2Iin_2Xx.....Buff.	2....$C_{ii}In_2Xx$.....Buff.
1....$C_2In_2x_2$.....White.	2....$C_2Iin_2x_2$.....White.	1....$C_{ii}In_2x_2$.....White.

Uniting the blacks and black-and-reds (since red appears only in one sex and often not until late in life) we find the following relation between the calculated and the observed proportions in 86 offspring: Calculated, black 65, buff 16, white 5; observed, black 61, buff 17, white 8.

In still another pen (848) the F_1 hybrids were mated to a Buff Cochin. Only 21 chicks were raised. Expectation is, black 10.4, buff 5.2, white 5.2. Actually there were obtained, black 7, buff 10, white 4. Half of the calculated blacks are really heterozygous in both black and buff; so expectation is a little uncertain, and probably should be given as something under 10.4. Also, on account of small numbers, a close agreement is not to be expected; but calculation and observation are at least of the same order.

CHAPTER XI.

INHERITANCE OF BLUE COLOR, SPANGLING, AND BARRING.

A. BLUE COLOR.

Color-patterns are generalized, like the barring, spangling, and "blueing"; or localized, like the wing-bar or hackle and saddle lacing. We have to consider at present the method of inheritance of the former of these kinds of color patterns. As is well known (Bateson, Saunders, and Punnett, 1902, 1903), the Blue or Andalusian fowl is a heterozygote and, as such, produces white gametes and also black gametes.* The "blue" is, indeed, a fine mosaic of white and black. The barbules of a blue feather are seen to be finely barred with alternating pigmented and unpigmented zones. The pigment consists of the ordinary melanic granules of a dark sepia color.

My original blues arose (in pen 502) from a White Leghorn hen B (recognized as heterozygous but of unknown origin), mated to a black Minorca. These blues are referred to in my 1906 report. They were both females and were mated (in pen 636) to a white cock (No. 340) similarly derived. Of 49 offspring, 11, or over 22 per cent, were black and 78 per cent either pure white (35 per cent of all), white with black specks (22.5 per cent) or white-and-black mosaic, *i. e.*, blue (20.4 per cent), but the latter were very variable in the quality of the blue. Let us designate the whitening factor of the White Leghorn by W (its absence w, resulting in black) and the blueing by M (its absence by m). Then, assuming that the blue females produce germ-cells MW, Mw, mW, mw, in equal numbers, and that the white male produces the same, we may expect in 16 F_2 offspring the combinations shown in table 64.

TABLE 64.—*Combinations in zygotes of the second hybrid generation of the blue strain.*

M_2W_2......1 white.	MmW_2.....2 white.	m_2W_2......1 white.
M_2Ww.....2 blue.	$MmWw$....4 white.	m_2Ww.....2 white.
M_2w_2......1 black.	Mmw_2......2 black.	m_2w_2......1 black.

Totals: White ten-sixteenths; black four-sixteenths; blue two-sixteenths.

The relation between the calculated and the actual percentages is as follows:

White + black specks in females: calculated, 62.5; actual, 57.5.
Black: calculated, 25; actual, 22.1.
Blue: calculated, 12.5; actual, 20.4.

---

*Wright (1902, p. 401) recognizes the variability of the blues. He advises the breeder of Andalusians that: "Black and white ones [offspring] can be weeded out at once; two or three months later birds absolutely too light, or dark and smoky, can be selected."

That the agreement is not closer must be attributed to the fact of small numbers and the premature death of many of the chicks, in consequence of which their adult plumage colors were not fully revealed. Also, many "blue" chicks produce white adults with black specks in the plumage.

It is to be observed that this explanation calls for a special mosaic (blueing) factor, but this mosaic factor brings about a blue plumage only when the "white" factor is diluted, *i. e.*, heterozygous.

In the next generation (pen 733) I mated 2 blues together. This mating is generally regarded as a unifactorial one (producing gametes WM, wM) and to give in every 4 offspring 1 black, 2 blue, and 1 white. I obtained the expected 50 per cent of blues, but always an excess of blacks and a deficiency of whites (49 : 35 : 16, respectively). This result is doubtless due to the accident that a large proportion of the chicks were described young, for it appears from my records that some blues become white when older and some "blacks" are certainly *blue-blacks*. The deficiency of whites becomes an *excess* of whites in the adult stage. The whites obtained from the blues are usually, but not always, splashed with black spots.

B. SPANGLING.

As is well known, hybrids between black fowl and White Leghorns are usually white with black patches in the females, while their brothers are mostly entirely white. This "spangled" condition is a heterozygous one just as truly as the "blue" condition is. When a splashed hen is mated to her white brother a certain proportion of the offspring are splashed again, *i. e.*, one-half of 50 per cent or 25 per cent, that being the proportion of heterozygous females. Actually in 150 offspring 19.4 per cent were splashed and 18.6 per cent black, while 62 per cent were recorded (largely from unhatched chicks) as pure white. The splashing reappears in about the expected proportion of cases. In my pen 633 I take the spangled females to form gametes WS, Ws, wS, ws, while the male seems to form gametes Ws, ws; S being the spangling factor. Then [♀ WS, Ws, wS, ws] × [♂ Ws, ws] gives the combinations shown in table 65.

TABLE 65.—*Combinations in zygotes of the second hybrid generation of the spangled strain.*

Zygotic formulæ.	Male.	Female.	Both sexes.
WSs	White	Spangled.	
Wss	White	White	
$2WwSs$	White, spangled.	Spangled.	
$2Wws$	White	White.	
wSs	Black.	Black.	
wss	Black.	Black.	
Total patterns in progeny:			
White	Five-eighths	Three-eighths	Eight-sixteenths.
Spangled	One-eighth	Three-eighths	Four-sixteenths.
Black	Two-eighths	Four-sixteenths	Do.

This analysis indicates that we should occasionally see a spangled male, and this expectation is realized. Thus No. 1250 ♂ is an F_2 out of White Leghorn A and the Rose-Combed Black Minorca No. 9. He is white with black spots covering about 10 per cent of the plumage, and No. 4222 ♂ of similar origin has much black on his chiefly white plumage. When they are mated to spangled hens of similar origin with themselves (pen 775), whites, blacks, and spotted, spangled, and blues occur in the proportions of 1, 17, and 12, respectively. Here again there is a deficiency of whites in the birds as described, a deficiency again probably due to immaturity.

Of the mottled condition all degrees are found, from white splashed with black to black with white spots; also, blue is very common in the offspring of two mottled birds. The relation of these patterns is very complex and much time would be required for their complete analysis, but it seems certain that there is a spangling or mottling factor, but that, as in canaries, guinea-pigs, and rats, the precise pattern is not inherited. There are, to be sure, in poultry, so called *races* of spangled birds with well-defined patterns, such as the spangled Polish, spangled Hamburghs, and so forth, but it is the experience of breeders that they do not reproduce their patterns closely. The prize-winning birds—those which conform to the breeder's ideals—are only a small proportion of each family of offspring. For instance, the Ancona type of plumage, which is black, each feather tipped with white, has to be carefully sought for in the progeny of each Ancona pen. The same is true of the Silver Spangled and Golden Spangled Hamburgs. There is little true spangling in the first plumage; the darker chicks prove the best; that is, there is the same tendency to grow whiter with age that I have noted above. And, finally, only a few birds in any flock are even fairly good show birds.

C. BARRING.

The presence of bands of black running at intervals across the otherwise white feather is a condition found in many types of poultry as well as various wild birds. It has become a fixed character in the Barred Plymouth Rock, which derived it in turn from the barred Dominique, whose barring was probably derived from the Cuckoo birds of England. Barring is also said to result from some crosses between white and black birds.

In my breedings barred birds have arisen several times:

(1) *White Cochin × Tosa.*—This case was referred to in my earlier report.* In the first generation of hybrids all males were barred. In the second hybrid generation I got 15 chicks that were white or nearly so, 25 with the Game color, and 16 barred. Remembering that only the males are barred and that the young heterozygous females are classed with Games, it appears that the barring is a heterozygous condition, occurring actually or potentially in about 50 per cent of the second hybrid generation

---

* 1906, page 49, figs. 35, 37, 37a.

and that the whites and some of the Games are extracted types. This conclusion is confirmed by further breeding. In pen 663 I bred 2 extracted white hens of Cochin-Tosa origin to a white cock and got 12 chicks, of which all were white, except that 3 showed a trace of reddish color. From the extracted Games bred together I got 36 chicks, all Games. In the case of this cross, consequently, barring is clearly heterozygous and confined to the male sex.*

(2) *White Leghorn Bantam × Dark Brahma.*—This cross was referred to in my report of 1906. From the table given there it appears that I got 5 barred fowl in F_1 out of a total of 51. In pen 701 I attempted to see if I could fix this barring. I used the best barred cock of the F_2 generation and the best barred hens of F_1 or F_2. The result was as shown in table 66.

TABLE 66.—*Distribution of color in F_2 or F_3 hybrids of the barred strain.*

[ABBREVIATIONS: W.L. = White Leghorn; Dk.Br. = Dark Brahma.]

Mother.				Father.				Offspring.			
No.	Gen.	Races.	Color.	No.	Gen.	Races.	Color.	White	Black.	Dark Brah.	Barred.
721	F_1	W.L. × Dk.Br.	Dark barred...	1898	F_2	W.L. × Dk.Br.	Barred.	....	5	7	5
894	F_2	....Do......	Well barred....	1898	F_2	....Do.......	..Do...	....	9	3	*10
965	F_2	....Do......	Medium barred	1898	F_2	....Do.......	..Do...	2	16	4	8
1335	F_2	....Do......	Dark barred...	1898	F_2	....Do.......	..Do...	1	14	1	2
1772	F_1	....Do......	Poorly barred..	1898	F_2	....Do.......	..Do...	....	4	7	†5
1915	F_2	....Do......	Fairly barred..	1898	F_2	....Do.......	..Do...	....	10	4	5
2576	F_2	....Do......	....Do.......	1898	F_2	....Do.......	..Do...	....	9	11	3
		Totals (145)............................						3	67	37	38
		Percentages.............................						2.1	46.2	25.5	26.2

*Including 1 blue.   †Including 2 blue.

This result suggests the interpretation that one of the parents, probably the male, contains both heterozygous black and barring, while the other parent lacks the supermelanic coat and has homozygous barring. Then of the offspring half will be barred and half will be black and, consequently (since only the non-black show their barring), one-fourth will appear barred, one-fourth will appear of the Dark Brahma type, and half will be pure black or have the pattern obscured by the supermelanic coat.

(3) *White Leghorn Bantam × Black Cochin.*—In still another experiment (pen 511) I crossed a White Leghorn bantam and a Black Cochin as described in my report of 1906. Of 24 hybrids that developed, 10 were white or nearly so, 7 were black, and 7 were barred black and white. The White Leghorn was heterozygous in white (half of the offspring being not white) and heterozygous to barring. In pen 650 the barred birds were mated together with results as given in table 67.

---

* Goodale, 1909, has shown that in Plymouth Rocks males may be and females usually are heterozygous in barring. There is thus a clear difference between the barring of the Cochin × Tosa hybrid and that of the Plymouth Rock. The question of the heterozygous nature of the female sex, fully discussed by Goodale, will be considered by me in another place. [Note at time of correcting proof.]

INHERITANCE OF BLUE COLOR, SPANGLING, AND BARRING. 83

On the assumption that the zygotic formula of both hens and cocks is BbN_2Ww (compatible with a barred plumage) we get four-sixteenths of the offspring white, three-sixteenths mottled or barred and nine-sixteenths black or Game, thus approximating the observed result; i.e., 21, 16, 47 as compared with 23, 21, 40. The result supports the hypothesis of a barring factor, B.

TABLE 67.—*Distribution of color in offspring of barred White Leghorn × Black Cochin hybrids.*

Mother.				Father.				Offspring.		
No.	Gen.	Races.	Color.	No.	Gen.	Races.	Color.	Wh.	Spangled, barred and blue.	Black or Game.
263	F_1	Bl. Coch.×Wh. Legh.	Barred..	205	F_2	Bl. Coch.×Wh. Legh.	Barred..	8	8	16
361	F_1	....Do...............	..Do....	205	F_2	....Do...............	..Do....	7	4	15
364	F_1	....Do...............	..Do....	265	F_2	....Do...............	..Do....	8	9	9
		Total...						23	21	40

CHAPTER XII.

GENERAL DISCUSSION.

A. RELATION OF HEREDITY AND ONTOGENY.

In studying heredity our attention must often be focused on the ontogenesis of the different characters, and we are sometimes inclined to regard the adult character as the product of the course of ontogenesis. But this is a superficial way of looking at things; the determiners of all characters are in the germ-plasm and together they direct the development of one part after another in orderly succession; a modernized form of the preformation doctrine seems logically necessary.

What do we know of the processes that take place in bringing the fertilized egg, freighted with its specific heredity, to its destination—the adult form? Modern embryological and cytological studies give us an insight into many of them. First of all, the egg has a certain organization that foreshadows something of its fate. Then cell-divisions begin, at first synchronous, but later becoming accelerated here and retarded there. Eventually (especially among animals) these cells become arranged into a membrane whose unequal growth in limited areas produces foldings. The folding of membranes, their stretching, local thickenings, or thinnings are largely the result of local inhibitions of water. Sometimes movements of individual cells occur out of the membranes into and through cavities or solid yolk-masses, and by the aggregation of such cells massive organs are sometimes formed. Local absorption of tissues already established may be effected in later life by such migratory cells. Membranes once established may form pockets or linear folds, as in gastrulation and gland formation; they may become perforated; two membranes may fuse along areas or lines and a perforation may even occur at the region of fusion. Linear strands or tubules may grow out, making connections, as nerves do, with distant organs; tubes may unite to form a network, or split lengthwise. Finally, membranes and masses undergo vacuolization, or masses may split apart or fuse together. Thus in the ontogeny that is proceeding under the control of heredity all is motion and change.

What are the factors that control all these movements—for these are the true factors of heredity? We do not know much about them, but we know some things. We know that cell-divisions occur at particular times and places under the influence of preceding division planes; but their normal occurrence may be interfered with by an abnormal chemical condition of the environment.

We have reason for concluding that each developmental process is a "response"—a reaction of the living, streaming protoplasm to changing environment. The nature of the response to any stimulus probably depends on the chemical constitution of the protoplasm—and this is hereditary. In an important sense heredity is the control of ontogeny.

The *specific* characteristics are mostly those that appear late in ontogeny. The integumentary folds over the nasal bones of the chick appear on or about the tenth day. At that time it can be ascertained whether the comb is median, or multiple, or Y-shaped, or cup-shaped, or consists of 2 papillæ. In the case of the single-comb the fold is linear and single; in the case of the pea-comb, linear and triple; in the case of the rose-comb, quintuple or irregularly wrinkled over the whole area; in the case of the Polish-comb, there is a pair of "pocket folds." In the single-combed fowl the single linear fold grows quickly to a great height and very thin, while in the pea-comb, with its additional pair of wrinkles, the median element is not so high as in typical single-combed races; in the pea-comb there is an additional folding stimulus and a reduced growth stimulus. In the heterozygote both stimuli are weakened; the lateral folds are usually much reduced—"are hard to make out," as I stated in 1906 (p. 35); and the factor that determines the continued growth (elevation) of the fold is weakened, so that the pea-comb—although "abnormally high" (1906, p. 35, figs. 20 and 21)—is not nearly as high as the single-comb of the Minorca (1906, fig. 4).

Two results are evident: first, each character in the heterozygous condition is reduced, and, second, each is much more variable than in the homozygous condition. Why is the character reduced? If the reaction to continued growth of the fold is strong in one race and weak in the other, then in the heterozygote that reaction, whatever its nature, is reduced. Why is the reduction in the response so variable? There is a variation in the irritability or other growing factor of the embryonic material that is destined to form the fold. Even Minorcas vary in the growth of the comb, and so do the Dark Brahmas. Let G be a constant element of the growth factor of the Minorca's comb; then $G + a$ or $G - a$ will indicate its variants. Let g be the growth factor of the Brahma's comb, and $g + a$ and $g - a$ its variants. Then the hybrids of these two races may be of the following types: Gg, $Gg + a$, $Gg - a$, $Gg + 2a$, $Gg - 2a$. This gives 5 varying conditions instead of 3 and greater extremes of variation.

In the foregoing case I have assumed that the positive character is that of increased growth in the Minorca; but the positive character may be an *inhibition* to indefinite growth of the pea-comb. Heredity may be conceived of as exerting at all points a control on developmental processes—sometimes initiating and continuing this; but often, on the other hand, slowing down or wholly inhibiting that. The inhibition of a process is quite as *positive* a function of heredity as its initiation. The hair of a young rabbit grows until it attains a certain length and then the growth ceases. The growing character is a youthful, embryonic one; the new character is the stoppage of growth. Similarly the young feathers of birds grow continuously until something intervenes that stops the growth and dries up the sheath. Now, in Angora rabbits and long-tailed fowl the epidermal organ continues its embryonic growth indefinitely; the something that intervenes to stop growth is absent. There is no reason for regarding the long hair or long feather as a positive condition and short hair or feather as due to its absence.

Again, Mediterranean fowl have non-feathered shanks; but in Asiatics the feet are feathered like the rest of the body (except the soles and face). It has been assumed that *boot* is an additional character and should be dominant over absence of boot. But, on the other hand, we may well think of the capacity of producing feathers as general to the skin. From this point of view the real question is, what prevents feather production on the eyelids, comb, wattles, and shank? It seems equally probable that there is an inhibitor of feather-growth for these few areas as that every conceivable area of the body has its special stimulus factor for feather development; or even as that there is such a factor to each separate feather-tract. In the Minorca, then, the inhibitor of boot is present; in the Silkie a weak heterozygous inhibition appears; but in the Dark Brahma there is no inhibitor and feathers extend down from the heel over the whole of front and sides of the foot and even on the upper surface of the toes—just as they do over the anterior appendages.

The case of the rumpless fowl is important in relation to the hypothesis of inhibitors. Either tail-production depends on a special factor TT, which is diluted, as Tt, in the heterozygote; or else there is a tail inhibitor, II, which is diluted, as Ii, in the heterozygote. In F_2 we expect, on the one hypothesis, 25 per cent tt, giving no tail, and 25 per cent TT, giving tail; on the other hypothesis 25 per cent ii, giving tail, and 25 per cent II, giving no tail. Actually we get all tailed in some cases; in others 25 per cent with no tail. Which hypothesis best fits the facts? Which is the more probable—that the 25 per cent recessive no-tail should produce a tail (as it were, out of nothing) or that the 25 per cent dominant tail inhibitor should be ineffective, permitting the development of a tail? It is clear that the ontogenetic failure of an inhibitor is easier to understand than the development of a character that is not represented at all in the germ-plasm. This matter is treated in another connection in the next section. But the present point is that it is equally in accord with the facts to regard heredity as initiating and inhibiting processes. If, indeed, processes were not regularly inhibited, they must, when once started, go on indefinitely, as do the hairs of Angora goats and wonder-horses.

As we have seen, ontogeny is not completed at hatching or birth. Many characters are at that time undeveloped. Hence, not infrequently the recessive condition is at first seen and is only later replaced by the dominant condition. The reverse sequence will rarely be followed, because development rarely, except in cases of degeneration, moves backward. One of the familiar cases of this sort is human hair-color. In youth this is frequently flaxen, later it becomes light brown, and eventually it may become dark brown. Darwin gives a number of examples in his Chapter XII of Animals and Plants under Domestication. To these I may add some from my own experience. The hybrids between white and gray Java sparrows are at first light and later become of a slaty gray like the dark parent. Many black fowl gain white feathers as they grow older, and every fancier knows that birds with complex white-and-black patterns can usually be "exhibited" only once, on account of loss of "standard" coloration late in life. In these cases the advanced condition in the series of melanic colors appears only late in ontogeny.* Similarly Lang (1908, p. 54) finds that in snail hybrids often the young shells have the recessive yellow color, only later in life showing the dominant red color. This is, of course, no reversal of dominance in ontogeny, but mere ontogenesis of pigmentation. So in general, since the recessive condition is absence of the character or its low stage of development and the dominant condition is presence of the full character, the individual in ontogenesis may exhibit in succession the recessive and then the dominant character, but not in the reverse order.

B. DOMINANCE AND RECESSIVENESS.

If segregation is the cornerstone of modern studies in heredity, dominance forms an important part, at least, of the foundation. In any case, a critical examination of dominance is now required; the more so since its significance and value have often been doubted.

First, how is a dominant character to be defined? It has been defined both on the basis of visible results in mating and on the basis of its essential nature. On the basis of visible results in hybridizing dominant characters may be defined as Mendel (1866, p. 11) defined them: "jene Merkmale, welche ganz oder fast unverändert in die Hybride-Verbindung übergehen." Bateson's translation (1902, p. 49) renders this passage: "those characters which are transmitted entire, or almost unchanged in the hybridization."

On the basis of the essential nature of the dominant character there has obtained a great diversity of definitions. Thus de Vries (1900, p. 85) suggested that the "systematically higher" character is the dominating one, and, again (1902, pp. 33, 145), that the dominant character is the phylogenetically older one. Many have suggested that it is the positive or present character that dominates over the negative, latent or absent. This last idea has become the prevailing one and its history is worth summarizing.

As early as 1902, Correns used as Mendelian pairs, presence of coloring material and absence; also modification into yellow and *no* modification.

---

* Does the graying of human hair represent an ontogenetically advanced condition of the melanic pigment as yellow represents the embryonic condition?

In 1905, he extended somewhat this use of present and absent characters, *k* (keine) preceding the symbol of a character as a negative. Still he did not pretend to generalize the relation of dominance and recessiveness to be that of presence and absence. In 1903 (p. 146) de Vries stated that in very many cases Mendel's law held when one quality is active and the other latent, and that the active quality is dominant. His illustrations show that by activity he meant essentially presence, by latency absence from the visible soma. Bateson's third report (1906) applies presence and absence to several additional cases, and, at the International Genetics Conference of that year, Hurst developed the presence-and-absence hypothesis, favoring the view that the factor for absence is nothing at all, but finding that certain cases, such as Angora coat, offer a difficulty. At the same meeting I suggested that "a variation * * * that is due to abbreviation of the ontogenetic process, which depends on something having dropped out, will be recessive," a progressive variation dominant; and in 1908 I expressed the conclusion that "dominance in heredity appears when a stronger determiner meets a weaker determiner in the germ. The extreme case is that in which a strong determiner meets a determiner so weak as to be practically absent, as when a red flower is crossed with white." I suggested that in some cases of recessiveness of an apparent advanced condition, like Angora hair, the dominant factor is an inhibitor. In the last year or two the presence-and-absence theory has gained wide acceptance, but I still think the cases where there is dominance of the advanced condition over the less advanced—of the quantitatively well-developed over the quantitatively less well-developed—have not been sufficiently considered. In human hair-color any other hypothesis demands that there are many units in the higher grades of pigmentation and fewer in the lower grades and that the presence of the surplus factor in any other higher grade dominates over its absence in the next lower grade; but there is no evidence in human hair-color of distinct, discontinuous units in the common yellow-brown series. And, in ontogeny, the different grades of color form a *continuous* series whose development proceeds throughout early life and may even be stimulated to an advanced stage of darkening by disease. The *cessation* of color development may take place at any point, and this seems incompatible with the theory of unit-characters for the different grades of human hair-color. In the present paper, on the other hand, the characters dealt with are mostly unit-characters and their quantitative variations mostly heterozygotic. Even the case of the Silkie boot (table 31, C) referred to in an earlier paper * as illustrating recessiveness of the less advanced condition proves, on further analysis, to be a case of heterozygotism. It seems highly probable that the future will show that many more advanced or progressive conditions are really due to one or more unit-characters not present in the less advanced condition. In that case it will appear that there is perfect accord in the two

---

* Davenport, 1908, page 60.

statements that the progressive condition and the "present" factor are dominant.

The definition of dominance on the ground of results meets at the outset with a difficulty the germ of which is observable in Mendel's cautious statement "ganz oder *fast* unverändert." Even Mendel observed that the hybrids between white-flowered and purple-red flowered peas have flowers less intensely colored than the darker parent. The experiments of the last seven years have shown that the "dominant" character is often very greatly changed—indeed, in extreme cases a blending of characters may occur—in the first generation. Correns (1900 b, p. 110) very early stated that in a certain set of crosses between good species the hybrids showed the character of *both* parents, only reduced, but in varying degrees. Bateson and Saunders (1902, p. 23) found in crossing two forms of *Datura* that—

> Although the offspring resulting from a cross between any two of the forms employed are usually indistinguishable from the type which is dominant as regards the particular character crossed, yet in other cases the intensity of a dominant character may be more or less diminished either in particular individuals or in particular parts of one individual. In *Tatula-Stramonium* cross-breds the corolla is often paler in color than that of the dominant parent (as has already been noticed by Naudin), but even in the palest specimens the deep blue color of the unopened anthers leaves no doubt as to the presence of the dominant color element. * * * The occurrence of intermediate forms was also occasionally noticeable in the fruits. Among the large number of capsules examined, there were some of the mosaic type, in which part of the capsule was prickly and the remainder smooth, while others, suggesting a blend, were more or less prickly all over, but the prickles were much reduced in size, and often formed mere tubercles.

Bateson and Saunders further showed (1902, p. 123) that in the case of comb and extra-toe in poultry "the cross-bred may show some blending and * * * the intensity of the dominant character is often considerably reduced."

Correns (1905, p. 9) pointed out that there was known, even at that time, a complete series of cases at one extreme of which one determiner completely hindered the appearance of the other, while at the opposite end of the series the hybrid showed an intermediate condition, both determiners appearing with equal strength.

The following year, in my first report on Inheritance in Poultry, I laid great stress on the imperfection of dominance, and this phenomenon has become more striking and clear in the subsequent years, until in the present paper it is recognized as the key to the explanation of many apparently anomalous types of heredity.

The first case in the present work in which imperfection of dominance is considered is that of the hybrids between I and oo comb. Here median comb is mated with no-median. Each somatic cell of the hybrid—at least in the comb region—has only half the full determiner for median comb. The determiner is weakened, and so the median comb is imperfectly developed, namely, at the anterior end of its proper territory. The weakening varies much in degree in the heterozygote. The median comb may be reduced to 70 per cent of its normal length or it may not develop at all.

The second case of imperfection of dominance is that of polydactylism. Extra-toe mated to normal gives extra-toe in 73 per cent only of the offspring in the case of the Houdans. Any trace of 6 toes (on one or both feet) is found in only 12 per cent of the hybrid offspring from a 6-toed Silkie parent. Certainly dominance here is very like blending.

The third case of imperfection of dominance is that of syndactylism. No syndactyls were noticed in F_1. My first conclusion was that syndactylism is recessive; but later studies have shown that it is dominant and that all matings of two syndactyl parents yield about 56 per cent syndactyl offspring.

Rumplessness gives an illustration of how dominance may be so weak as to be absent altogether; so that from F_1 alone the erroneous conclusion is drawn that it is recessive; indeed, in one strain, only faint traces of the character made their appearance in successive generations.

Finally, winglessness is a character which *appears* not to be inherited at all. Nevertheless our experience with rumplessness leads us to suspect that winglessness also is an impotently dominant character.

Looking at the matter frankly and without prejudice, the question must be answered: Has not the whole hypothesis of dominance become *reductio ad absurdum?* What visible criterion of dominance remains, where dominance fails completely? All the usual statistical landmarks of proportional appearance in successive generations being lost, can one properly speak of dominance and recessiveness at all?

Amid the general ruin of criteria, however, one means of detecting dominance remains. That extracted character which in F_2 or subsequent generations shows in homologous * matings in some families a wide range of variability is dominant, while that extracted character which constantly, in all homologous matings, shows no or very little variation is recessive.

The reason for this difference in the inheritableness of the two conditions is easy to understand on the principles enumerated in the last section. A positive character has a real ontogeny. But, as we have seen, the development of any character may be *interrupted* at any stage. Most aberrations among organisms are due to a retardation or failure of normal development. In human affairs we recognize this tendency in the terms "degenerates" and "defectives" (constituting from 2 to 4 per cent of the population). Indeed, there are few persons who are not defective in some physical or psychical character. In cases where the commonest form of abnormality is due to a development *in excess* it seems probable that a normal restraining or inhibiting factor is defective or absent. On page 88 I tried to show how common in ontogeny such restraining and inhibiting factors are. Since onto-

---

* By *homologous* matings I mean those in which the germ-plasms of *both* parents are in the same condition with reference to the unit-character; i. e., both either possess it pure or lack it altogether.

genetic processes are so often cut short by external conditions, we can understand the *variability* in the degree of development of positive characters.

On the other hand, whenever the fundamental hereditary stimulus or the material for a character is absent from the germ-plasm of both parents, then it can appear in none of the offspring; they will be practically invariable in respect to this condition. Only the ontogenetic fluctuations of other real characters may influence the defect. Consequently the absent state reproduces itself, the "recessive breeds true."

The considerations here presented bear upon the hypothesis of change of dominance. Bateson and Punnett (1905, p. 114) say of poultry: "The normal foot, though commonly recessive, may sometimes *dominate* the extra-toe character." This idea of occasional *change* in dominance has been expressed more than once in the literature. I think the phrase an unfortunate one. In my earlier report * I urged that a characteristic that is anywhere dominant is so without regard to race or species involved. If this is so it is clearly improbable that it should vary from individual to individual, or in the same individual at different times. Rather in view of the imperfection of dominance we should say that a dominant character sometimes fails to develop, in which case it is absent from the progeny; that is all. It is particularly apt to fail of development when dilute—*heterozygous*.

C. POTENCY.

Perhaps an apology is needed for introducing the much-abused word "potency"; but there is hardly another that can be so readily adapted to the precise definition I desire to give to it. The potency of a character may be defined as the capacity of its germinal determiner to complete its entire ontogeny. If we think of every character as being represented in the germ by a determiner, then we must recognize the fact that this determiner may sometimes develop fully, sometimes imperfectly, and sometimes not at all. When such a failure occurs in a normal strain a sport results.

Potency is variable. Even in a pure strain a determiner does not always develop fully, and this is an important cause of individual variability. But in a heterozygote potency is usually more or less reduced. When the reduction is slight *dominance* is nearly complete; but when the reduction is great dominance is more or less incomplete and, in the extreme case, may be absent altogether. The series of cases of varying perfection of dominance described in this work illustrate at the same time varying potency. The extreme case is that of the rumpless fowl. The character in this case is an inhibitor of tail development. This character has arisen among vertebrates repeatedly and has become perpetuated in some amphibia and primates, including man. In the case of our cock No. 117, the action of the inhibitor is very weak, so that in the heterozygote the development of the tail is not interfered with at all and even in extracted dominants it

---

* Davenport, 1906, page 86.

interferes little with tail development, so that it makes itself felt only in reduced size of the uropygium and in bent or shortened back. But in No. 116 the inhibiting determiner is strong. It develops fully in about 47 per cent of the heterozygotes and 2 extracted dominants may produce a family in *all* of which the tail's development is inhibited. In the case of the rumpless condition that arose apparently *de novo* in my yards, the new inhibitor showed an intermediate potency completely stopping the tail development in 1 out of 25 heterozygotes. These three cases afford a striking illustration of a variation in the potency of the same inhibiting character in different strains.

Not only is potency variable, but its variations seem, in some cases, to be inheritable. This we have seen to be the case with the Y-comb (p. 15); with the extra-toed condition of Houdans (p. 23); and with rumplessness (*cf.* offspring of No. 117 as compared with No. 116, p. 40). On the other hand, the extra-toed condition of Silkies, the grade of clean shank, and the degree of closure of nostril seem not to be inherited.

D. REVERSION AND THE FACTOR HYPOTHESIS.

The brilliant development of the factor hypothesis, only dimly foreshadowed by Mendel * (1866, p. 38), clearly expressed by Correns (1892), applied to animals by Cuénot, and further elaborated by Bateson and Castle and their pupils, has quite changed the methods of work in heredity. More forcibly than ever is it brought home to us that the constitution of the germ-plasm—not merely the somatic character—is the object of our investigation. With this principle fully grasped the existence of cryptomeres and the resolution of characters have become clearer. But the most striking result accomplished has been that of clearing up the whole range of phenomena formerly placed in the category of "reversion." No idea without a semblance of inductive explanation has been more generally accepted in the Darwinian sense both by professed biologists and practical breeders than this. Not only was the fact of recurrence of ancestral types in domesticated organisms accepted, but the idea that, in some way, hybridization *per se* destroyed the results of breeding under domestication was maintained.† Now we know that, under domestication, many races have been preserved that are characterized by a deficiency of a character or by a new, additional one, and that hybridization, by bringing together again those characters that are found in the ancestral species, may bring about again individuals of the ancestral type. There is nothing more mysterious about reversion, from the modern standpoint, than about forming a word from the proper combination of letters.

---

* Mendel's expression on this subject is translated by Bateson (1902, p. 84) as follows: "Whoever studies the coloration which results in ornamental plants from similar fertilisation can hardly escape the conviction that here also the development follows a definite law which possibly finds its expression in *the combination of several independent color characters*. (The italics are Mendel's.)

†"An inherent tendency to reversion is evolved through some disturbance in the organisation caused by the act of crossing." (Darwin, Animals and Plants under Domestication, Chapter XIII, section, "Summary on proximate causes leading to reversion.")

E. THE LIMITS OF SELECTION.

In the last few decades the view has been widespread that characters can be built up from perhaps nothing at all by selecting in each generation the merely quantitative variation that goes farthest in the desired direction. I have made two tests of this view, using the plumage color of poultry.

(1) *Increasing the red in the Dark Brahma × Minorca cross.*—The Dark Brahma* belongs to the group of poultry that contains a majority of characters derived from the Aseel type. Nevertheless, its plumage is closely related to that of the Jungle-fowl, from which it may be derived on the assumption that the red part of the pattern has become, for the most part, white. However, a little red remains on the middle of the upper feathers of the wing-bar. I crossed such a bird with a Black Minorca, and, as reported in my earlier work,† the offspring were all black, except that the males showed some red on the wing-bar. The amount of red varied in the different males, and I decided to test the possibility of much increasing the amount of the red by selection in successive generations. So I chose the reddest cock to head the pen. In this pen (No. 632) 222 chicks were produced and grew to a stage in which their adult color could be determined. Of these 222 chicks, 160, or 72 per cent, were black, without red; 24, or 10.8 per cent, were black with some red; 38, or 11.7 per cent, were typical Dark Brahmas, and 9 others, or 4.5 per cent, were modified Dark Brahmas.

The following year (pen 732) I bred a cock derived from the last year's pen, a bird that resembled much the male Dark Brahma (except that it was somewhat darker), to sundry hens, hybrids between the Dark Brahma and Minorca—some of the first and some of a later hybrid generation, but all black except that some of the 1906 birds had a little buff on the breast and the primaries. The F_1 (black) $\times F_2$ (Dark Brahma) gave 51 per cent black offspring, 27 per cent with a black-and-red Game pattern, and 22 per cent with the Dark Brahma pattern devoid of red. Thus the third generation suddenly gave me a red-and-black Game-colored bird (plate 12)!

My interpretation of the foregoing results is as follows: The Dark Brahma gametic formula proves to be $CIrnwx$, whereas the Black Minorca is $C(IR)Nwx$, where (IR) is equivalent to, and merely a further analysis of, the J of the formula of the Minorca as given in earlier sections. The I stands for the Jungle pattern without red and R is the red element in that pattern. Obviously N and R are the differential factors, 4 kinds of gametes occur in F_1, and in every 16 offspring these factors are combined in the following proportions: 9 NR, 3 Nr, 3 nR, 1 nr (compare the distribution of color types in the 222 offspring of pen 632). The F_2 male selected as father of the next generation (in pen 732) was an extracted Dark Brahma in coloration and probably formed only 1 kind of gamete, nr; but the hens

---

* Plate 11.   † Davenport, 1906, page 35.

were heterozygous in respect to N and R. Consequently 4 kinds of zygotes are to be expected in F_2; and expectation was realized as indicated in table 68.

TABLE 68.

	$NnRr$. Black with traces of red in male.	$Nnrr$. Black.	$nnRr$ Game.	$nnrr$. Dark Brahma (without red).
	P. ct.	P. ct.	P. ct.	P. ct.
Expectation	50	25		25
Realization	51	27		22

In the case where both parents are F_2 or F_3 it is impossible to summate results, since the gametic formulæ of the different parents are so diverse; but the same types of solid blacks, black with trace of red in the males, Game-colored males and females, and Game with red replaced by white repeatedly occur. My plan of increasing red in the Dark Brahmas met with wholly unexpectedly prompt success, but not in the way anticipated. The result was not due to selection, but to the recombination of the factors necessary to make the Game plumage coloration.

(2) *Production of a buff race by selection.*—The second test was directed toward the production *de novo* of a new buff race from a Game fowl.

As is well known, all of our red and "buff" races, like the Buff Leghorn, Rhode Island Red, and others, have been derived from the Buff Cochin that came to us from China. The fact that a buff bird has, so far as I have been able to learn, not been produced in western countries indicates the probability that it can not be so produced at will; but the attempt seemed worth while.

I began with a Black Breasted Red Game because its plumage color is that of the primitive ancestor of domesticated poultry, and on that hypothesis the ancestor of the buff races. If these buff races were produced by extending the red through selection of the reddest offspring, that should be possible now as in the past.

A start in the direction of creating a buff bird would seem to require the elimination of the black. By crossing a black and red Game with a White Leghorn I got, in 1905, 2 white pullets with red on breast and some black specks. By crossing a Game Bantam (wingless) with a White Leghorn I got white birds with red present on wing-bar of male and breast of females and also some black spots.

In 1906 I mated 2 of these white (+ red) bantam hybrid hens with a hybrid cock and obtained again red on the wing-coverts of some white hybrids, while some were without red. From one of the hens I got 4 offspring, or 20 per cent of all, with buff on hackle-lacing, breast, and wing-coverts.

In 1907 I mated a prevailingly white male of the preceding year, that had red wing-bar, hackle, and breast, with the reddest females and obtained, along with pure whites and blacks and barred birds, these colors combined with red in various degrees, but not clearly in advance of the reddest of 1906. In 1908 I mated a white male, having red as in the Game, with my reddest hybrids. Again, white and white-and-buff birds appeared, but they showed no advance, except in one instance, among 138 young. This individual (No. 7950), derived exclusively from the Black-red Game and White Leghorn on one side and on the other from the White Leghorn-Game Bantam cross, had a *uniform buff* down. Unfortunately the chick quickly died.

The conclusion is that after three years of selection of the reddest offspring no appreciable increase of the red was observed—except for the remarkable case of one undeveloped chick with completely buff down. This, indeed, looks like a sport, or, perhaps, it is due to unsuspected factors. The experiment will be continued.

F. NON-INHERITABLE CHARACTERS.

So well-nigh universal is heredity that it is justifiable to entertain a doubt whether any character may fail of inheritance. So far as my experience goes, non-inheritable characters are such as are weak in ontogeny, so that they may readily fail of development even when conditions are propitious; or else they are so complex—so far removed from simple unit-characters—that their heritability in accordance with established canons is obscured. The first case is apparently illustrated by the rumpless cock (No. 117) and the wingless fowl; the second case by lop-comb and by right-and-left alternatives in general.

Apart from the distinct *characters* that fall under these two categories there are the fluctuating quantitative *conditions*. These depend for the most part, as already pointed out, on variations in the point at which the ontogeny of a character is stopped; and the stopping-point is, in turn, often, if not usually, determined by external conditions which favor or restrict the ontogeny. Whether or not such quantitative variations are transmitted is still doubtful. Our experiment in increasing qualities, such as redness in plumage-color, by selection of quantitative fluctuations have not been successful in the sense anticipated; neither have selections of comb, polydactylism, or syndactylism. Recently, prolonged attempts at the Maine Agricultural Experiment Station to increase egg-yield of poultry by selection have been without result. Apparently, within limits, these quantitative variations have so exclusively an ontogenetic signification that they are not reproduced so long, at least, as environmental conditions are not allowed to vary widely.

The conclusions which others have reached, and upon which de Vries has laid the greatest stress, that quantitative and qualitative characters differ fundamentally in their heritability is supported by our experiments.

G. THE RÔLE OF HYBRIDIZATION IN EVOLUTION.

The criticism has often been made of modern studies in hybridization that they are really unimportant for evolution because hybridization is uncommon in nature. Even at the beginning of the new era it could be replied that, first, we did not know how common hybridization might turn out to be in nature, and, second, that certainly in human marriage and among domesticated animals and plants, intermixing of characters played a most important part, and, finally, the laws of inheritance of characters were of such grave physiological import as to deserve study wholly apart from any question of the rôle of hybridization in evolution.

The last decade of work has made clear many things that were before uncertain. We now realize that in nature hybridization may and actually does proceed extensively. Dr. Ezra Brainerd has shown how many wild "species" of *Viola* have arisen by hybridization, as may be proved by extracting from them combinations of characters that are found in the species that are undoubtedly ancestral to them. In such highly variable animals as *Helix nemoralis* and *Helix hortensis* it is very probable that individuals with dissimilar characters regularly mate in nature and transmit diverse combinations of characters to their progeny. Indeed, if one examines a table of species of a genus or of varieties of a species one is struck by the paucity of distinctive characters. The way in which species, as found in nature, are made up of different combinations of the same characters is illustrated by the following example, taken almost at random. Among the earwigs is the genus *Opisthocosmia*, of which the 5 species known from Sumatra alone may be considered. They differ, among other qualities, chiefly in the following characters (Bormans and Kraus, 1900):

Size: A, large; a, small.
Wing-scale: B, brown; b, yellow.
Antennal joints: C, unlike in color; c, uniform.
Forceps at base: D, separated; d, not separated.
Edge of forceps: E, toothed; e, not toothed.
Fourth and fifth abdominal segments: F, granular; f, not granular.

The combinations of these characters that are found are as follows:

Opisthocosmia ornata: $A b c D E F$.
insignis: $A B c D E f$.
longipes: $A b C D E f$.
tenella: $A b C d e f$.
minuscula: $a B C D E f$.

Other species occur, in other countries, showing a different combination of characters, and there are characters not contained in this list, which is purposely reduced to a simple form; but the same principles apply generally.

The bearing upon evolution of the fact that species are varying combinations of relatively few characters is most important. Combined with the fact of hybridization it indicates that the main problem of evolution is that of the origin of specific characteristics. A character, once arisen in an individual, may become a part of any species with which that individual can hybridize. Given the successive origin of the characters A, B, C, D, E, F, in various individuals capable of intergenerating with the mass of the species, it is clear that such characters would in time become similarly combined on many individuals; and the similar individuals, taken together, would constitute a new species. The adjustment of the species would be perfected by the elimination of such combinations as were disadvantageous.

COLD SPRING HARBOR, NEW YORK,
May 20, 1909.

LITERATURE CITED.

BALDAMUS, A. C. E.
    1896. Illustrirtes Handbuch der Federviehzucht. Erster Band: Die Hühnervogel. 3 Aufl. bearbeitet von O. Grünhaldt. Dresden, 1896. xvi+476 pp., 102 figs.

BARFURTH, D.
    1908. Experimentelle Untersuchung über die Vererbung der Hyperdactylie bei Hühnern. I. Mitth. Der Einfluss der Mutter. Arch. f. Entw.-Mech. der Org., xxvi, 631–650.

BATESON, W.
    1894. Material for the Study of Variation Treated with Especial Regard to Discontinuity in the Origin of Species. London, 1894, xvi+598 pp.
    1902. Mendel's Principles of Heredity: A Defence. Cambridge (Engl.), 1902, xv+212 pp.

BATESON, W., and SAUNDERS, Miss E. R.
    1902. Report I to the Evolution Committee of the Royal Society. London, 160 pp.

BATESON, W., and PUNNETT, R. C.
    1905. Experimental Studies in the Physiology of Heredity—Poultry. Report II to the Evolution Committee of the Royal Society. pp. 99–131.

BATESON, W., E. R. SAUNDERS, and R. C. PUNNETT.
    1906. Report III to the Evolution Committee of the Royal Society, London, 53 pp.
    1908. Report IV to the Evolution Committee of the Royal Society, London, 60 pp.

BORMANS, A. DE, and KRAUS, H.
    1900. Forficulidæ und Hemimeridæ. Das Tierreich, 11 Lief. Berlin, xv+142 pp.

CASTLE, W. E.
    1906. The Origin of a Polydactylous Race of Guinea-Pigs. Carnegie Institution of Washington Publication No. 49 (Papers of the Station for Experimental Evolution at Cold Spring Harbor, N. Y., No. 5).

CASTLE, W. E., MULLENIX, H. E., and COBB, S.
    1909. Studies of Inheritance in Rabbits. Carnegie Institution of Washington Publication No. 114 (Papers of the Station for Experimental Evolution No. 13).

CORRENS, C.
    1900a. G. Mendel's Regel über das Verhalten der Nachkommenschaft der Rassenbastarde. Ber. d. D. Bot. Ges., xviii, 158–167.
    1900b. Ueber Levkojenbastarde. Zur Kenntnis der Grenzen der Mendel'schen Regeln. Bot. Centralblatt, lxxxiv, 97–113. [Oct. 17].
    1902. Ueber Bastardirungsversuche mit Mirabilis-Sippen. Ber. d. D. Bot. Ges., xx, 594–608.
    1905. Zur Kenntnis der scheinbar neuen Merkmale der Bastarde. Ber. d. D. Bot. Ges., xxiii, 70–85.
    1905. Über Vererbungsgesetze. Verhandl. Ges. D. Naturf. u. Ärzte. Allg. Teil., 23 pp.

CUÉNOT, L.
    1903. L'hérédité de la pigmentation chez les souris (2me note). Arch. de zool. expér. et gén. (4), I. Notes et rev., pp. xxxiii–xli.

DARWIN, C.
    1876. The Variation of Animals and Plants under Domestication. Second edition, revised, vols. I, II. New York.

DAVENPORT, C. B.
    1906. Inheritance in Poultry. Carnegie Institution of Washington Publication No. 52 (Papers of the Station for Experimental Evolution, No. 7), v+136 pp., 17 plates.
    1907. Heredity and Mendel's Law. Proc. Washington Acad. of Sciences, ix, pp. 179–187 (July 31).
    1908. Determination of Dominance in Mendelian Inheritance. Proc. Amer. Philos. Soc., xlvii, 59–63 (April).

DAVENPORT, GERTRUDE C., and C. B.
    1909. Heredity of Hair Color in Man. Amer. Nat., xliii, 193–211 (April, 1909).

DE VRIES, H.
- 1900. Das Spaltungsgesetz der Bastarde. Ber. der D. Bot. Ges., XVIII, 83-90 (14 March).
- 1902. Die Mutationstheorie. Versuche und Beobachtungen über die Entstehung der Arten im Pflanzenreich, Zweiter Band. 1 Lieferung, pp. 1-240.
- 1905. Species and Varieties: Their Origin by Mutation. Ed. by D. T. MacDougal. Chicago, 1905 xviii+847 pp.

GOODALE, H. D.
- 1909. Sex and its Relation to the Barring Factor in Poultry. Science, XXIX, pp. 1004-1005. June 25.

HARRISON, R. G.
- 1901. On the Occurrence of Tails in Man, with a Description of the Case reported by Dr. Watson. Proc. 14th Ann. Sess. Assoc. Amer. Anat., p. 141-158, 5 pls.

HURST, C. C.
- 1905. Experiments with Poultry. In Report II to the Evolution Committee of the Royal Society (by Bateson et al.). London, 154 pp.

KEIBEL, F., and ABRAHAM, K.
- 1900. Normentafeln zur Entwicklungsgeschichte der Wirbelthiere. 2 Heft. Normentafel zur Entwicklungsgeschichte des Huhnes (Gallus domesticus). Jena, 1900, 132 pp., 3 Taf.

LANG, A.
- 1908. Ueber die Bastarde von Helix hortensis Müller und Helix nemoralis L. Eine Untersuchung zur experimentellen Vererbungslehre. Jena, 120 pp., 4 Taf.

LARRABEE, A. P.
- 1906. The Optic Chiasma of Teleosts: A Study of Inheritance. Proc. Amer. Acad. Arts and Sciences, XLII, 217-231 (Oct., 1906).

LEWIS, T., and EMBLETON, D.
- 1908. Split-hand and Split-foot Deformities, their Types, Origin, and Transmission. Biometrika, VI, 26-58, pls. I-VII (March).

LUTZ, F. E.
- 1908. The Inheritance of the Manner of Clasping the Hands. Amer. Nat., XLII, 195, 196 (March).

MENDEL, G.
- 1866. Versuche über Pflanzen-Hybriden. Verhandlungen des naturforschen Vereines in Brünn. Bd. IV, pp. 47.

SCHWALBE, E.
- 1906. Die Morphologie der Missbildungen des Menschen und der Tiere. 1. Th. Allgemeine Missbildungslehre. Jena, xvi+230 pp.

SPILLMAN, W. J.
- 1909. The Nature of "Unit" Characters. Amer. Nat., XLIII, 243-248 (April).

WRIGHT, L.
- 1902. The New Book of Poultry. London, etc., viii+600 pp.

PLATE 1

Jungle Fowl, male, showing distribution of black and red elements of pattern.

Jungle Fowl, female, showing coloration and pattern

PLATE 3

White-faced Black Spanish, male.

PLATE 4

PLATE 5

First generation hybrid between Black Minorca Cock and White Silkie Hen

Second hybrid generation between Silkie and Spanish Minorca. (No. 3898) female

Buff Cochin. (No 545) male.

PLATE 8

Cock of first hybrid generation between Black Cochin and Buff Cochin.

PLATE 9

Cockerel (No 6094) of first hybrid generation between Buff Cochin Cock and Silkie Hen.

PLATE 10

Cockerel (No. 2561) of second hybrid generation between Buff Cochin and White Leghorn

Dark Brahma. (No. 122) male.
The detailed feathers are in order from right to left from first, third and fourth wing coverts

PLATE 12

A cock (No. 5257) of the third hybrid generation between a single-comb Black Minorca and a Dark Brahma shown in plate 6.
The detailed feathers are in order from right to left from the first, second, fourth and third wing coverts.

Kenji Toda, pinx.